Radar Technology: Principles and Applications

Radar Technology: Principles and Applications

Edited by
Robert Campbell

Willford Press
www.willfordpress.com

Published by Willford Press,
118-35 Queens Blvd., Suite 400,
Forest Hills, NY 11375, USA

ISBN: 978-1-68285-487-7

Cataloging-in-Publication Data

Radar technology : principles and applications / edited by Robert Campbell.
p. cm.
Includes bibliographical references and index.
ISBN 978-1-68285-487-7
1. Radar. 2. Radar--Equipment and supplies. 3. Electronic systems. I. Campbell, Robert.

TK6575 .R33 2018
621.384 8--dc23

For information on all Willford Press publications
visit our website at www.willfordpress.com

Willford Press

Printed in China.

Contents

Preface

The object-detection system, which is used to find the velocity, range, and angle of the objects with the help of radio waves is called the radar. The detection of weather formations, aircrafts, guided missiles, terrain, spacecrafts, motor vehicles, etc. is made easy by radar technology. Some of the methods and principles used in this technology are Doppler effect, polarization, illumination, reflection, etc. The book studies, analyses and upholds the pillars of radar technology and its utmost significance in modern times. The topics included in it are of utmost significance and bound to provide incredible insights to readers. Those in search of information to further their knowledge will be greatly assisted by this textbook.

A detailed account of the significant topics covered in this book is provided below:

Chapter 1- Radars are used to determine the range and velocity of objects. Radar technology uses radio waves to detect objects. It is used in detecting guided missiles, ships, spacecrafts, aircrafts and terrain. This is an introductory chapter which will introduce briefly all the significant aspects of radar.

Chapter 2- To develop a better understanding of radar systems, it is important to understand monopulse radar, passive radar, pulse-Doppler radar, space-based radar and radar tracker. The pulse-Doppler radar uses the Doppler effect and provides detection of objects below a flying aircraft. This chapter is an overview of the subject matter incorporating all the major aspects of weather radars.

Chapter 3- Weather radar is a type of radar that is used in forecasting the weather. They are mostly pulse-Doppler radars that can detect the motion of rain droplets. The main types of radar output explained are plan position indicator, vertical composite, accumulations, echotops, range height indicator and automatic algorithms. This chapter will provide an integrated understanding of weather radars.

Chapter 4- The technological aspects discussed in this section are radar altimeter, radar detector, terrain-following radar, precision approach radar and radar warning receiver. Radar altimeter is used on aircrafts to measure the terrain beneath the aircraft whereas a radar detector is used in motorcycles to detect its speed. The aspects elucidated in this chapter are of vital importance, and provide a better understanding of radar systems.

It gives me an immense pleasure to thank our entire team for their efforts. Finally in the end, I would like to thank my family and colleagues who have been a great source of inspiration and support.

Editor

1

An Overview of Radar and its Types

Radars are used to determine the range and velocity of objects. Radar technology uses radio waves to detect objects. It is used in detecting guided missiles, ships, spacecrafts, aircrafts and terrain. This is an introductory chapter which will introduce briefly all the significant aspects of radar.

Radar

Long-range radar antenna, used to track space objects and ballistic missiles.

Radar is an object-detection system that uses radio waves to determine the range, angle, or velocity of objects. It can be used to detect aircraft, ships, spacecraft, guided missiles, motor vehicles, weather formations, and terrain. A radar system consists of a transmitter producing electromagnetic waves in the radio or microwaves domain, a transmitting antenna, a receiving antenna (often the same antenna is used for transmitting and receiving) and a receiver and processor to determine properties of the object(s). Radio waves (pulsed or continuous) from the transmitter reflect off the object and return to the receiver, giving information about the object's location and speed.

Radar was developed secretly for military use by several nations in the period before and during World War II. The term *RADAR* was coined in 1940 by the United States Navy as an acronym for RAdio Detection And Ranging or RAdio Direction And Ranging. The term *radar* has since entered English and other languages as a common noun, losing all capitalization.

The modern uses of radar are highly diverse, including air and terrestrial traffic control, radar astronomy, air-defence systems, antimissile systems, marine radars to locate landmarks and other ships, aircraft anticollision systems, ocean surveillance systems, outer space surveillance and rendezvous systems, meteorological precipitation monitoring, altimetry and flight control systems, guided missile target locating systems, ground-penetrating radar for geological observations, and range-controlled radar for public health surveillance. High tech radar systems are associated with digital signal processing, machine learning and are capable of extracting useful information from very high noise levels.

Radar of the type used for detection of aircraft. It rotates steadily, sweeping the airspace with a narrow beam.

Other systems similar to radar make use of other parts of the electromagnetic spectrum. One example is "lidar", which uses ultraviolet, visible, or near infrared light from lasers rather than radio waves.

History

First Experiments

As early as 1886, German physicist Heinrich Hertz showed that radio waves could be reflected from solid objects. In 1895, Alexander Popov, a physics instructor at the Imperial Russian Navy school in Kronstadt, developed an apparatus using a coherer tube for detecting distant lightning strikes. The next year, he added a spark-gap transmitter. In 1897, while testing this equipment for communicating between two ships in the Baltic Sea, he took note of an interference beat caused by the passage of a third vessel. In his report, Popov wrote that this phenomenon might be used for detecting objects, but he did nothing more with this observation.

The German inventor Christian Hülsmeyer was the first to use radio waves to detect "the presence of distant metallic objects". In 1904, he demonstrated the feasibility of detecting a ship in dense fog, but not its distance from the transmitter. He obtained a patent for his detection device in April 1904 and later a patent for a related amendment for estimating the distance to the ship. He also got a British patent on September 23, 1904 for a full radar system, that he called a *telemobiloscope*. It operated on a 50 cm wavelength and the pulsed radar signal was created via a spark-gap. His system already used the classic antenna setup of horn antenna with parabolic reflector and was presented to German military officials in practical tests in Cologne and Rotterdam harbour but was rejected.

In 1915, Robert Watson-Watt used radio technology to provide advance warning to airmen. During the 1920s, the UK research establishment, led by Watson-Watt, made many advances using radio techniques, including the probing of the ionosphere and the detection of lightning at long distances. Watson-Watt became an expert on the use of radio direction finding as part of his lightning experiments. As part of ongoing experiments, he asked the "new boy", Arnold Frederic Wilkins, to find a receiver suitable for use with shortwave transmissions. Wilkins made an extensive study of available units before selecting a receiver model from the General Post Office. Its instruction manual noted that there was "fading" (the common term for interference at the time) when aircraft flew by.

In 1922, A. Hoyt Taylor and Leo C. Young, researchers working with the U.S. Navy, had a transmitter and a receiver on opposite sides of the Potomac River and discovered that a ship passing through the beam path caused the received signal to fade in and out. Taylor submitted a report, suggesting that this might be used to detect the presence of ships in low visibility, but the Navy did not immediately continue the work. Eight years later, Lawrence A. Hyland at the Naval Research Laboratory observed similar fading effects from a passing aircraft; this led to a patent application as well as a proposal for serious work at the NRL (Taylor and Young were then at this laboratory) on radio-echo signals from moving targets.

Just before World War II

Experimental radar antenna, US Naval Research Laboratory, Anacostia, D. C., late 1930s

Before the Second World War, researchers in the United Kingdom, France, Germany, Italy, Japan, the Netherlands, the Soviet Union, and the United States, independently and in great secrecy, developed technologies that led to the modern version of radar. Australia, Canada, New Zealand, and South Africa followed prewar Great Britain, and Hungary had similar developments during the war.

In France in 1934, following systematic studies on the Split Anode Magnetron, the research branch of the Compagnie Générale de Télégraphie Sans Fil (CSF), headed by Maurice Ponte, with Henri Gutton, Sylvain Berline, and M. Hugon, began developing an obstacle-locating radio apparatus, a part of which was installed on the liner *Normandie* in 1935.

During the same time, the Soviet military engineer P. K. Oshchepkov, in collaboration with Leningrad Electrophysical Institute, produced an experimental apparatus, RAPID, capable of detecting an aircraft within 3 km of a receiver. The Soviets produced their first mass production radars RUS-1 and RUS-2 Redut in 1939 but further development was slowed by the NKVD arrest of Oshchepkov and their sending him to the gulag. In total, only 607 Redut stations were produced during the war. The first Russian airborne radar, Gneiss-2, entered into service in June 1943 on Pe-2 fighters.

More than 230 Gneiss-2 stations were produced by the end of 1944. The French and Soviet systems, however, had continuous-wave operation and could not give the full performance that was ultimately at the center of modern radar.

The first workable unit built by Robert Watson-Watt and his team

Full radar evolved as a pulsed system, and the first such elementary apparatus was demonstrated in December 1934 by the American Robert M. Page, working at the Naval Research Laboratory. The following year, the United States Army successfully tested a primitive surface-to-surface radar to aim coastal battery searchlights at night. This was followed by a pulsed system demonstrated in May 1935 by Rudolf Kühnhold and the firm GEMA in Germany and then one in June 1935 by an Air Ministry team led by Robert A. Watson-Watt in Great Britain.

A Chain Home tower in Great Baddow, Essex, United Kingdom

In 1935, Watt was asked to pass judgement on recent reports of a German radio-based death ray and turned the request over to Wilkins. Wilkins returned a set of calculations demonstrating the system was basically impossible. When Watt then asked what might they do, Wilkins recalled the earlier report about aircraft causing radio interference. This led to the Daventry Experiment of 26 February 1935, using a powerful BBC shortwave transmitter as the source and their GPO receiver set up in a field while a bomber flew around the site. When returns were clearly seen, Hugh Dowding, the Air Member for Supply and Research was very impressed with its potential and funds were immediately provided for development of an operational system. Watt's team patented the device in GB593017.

Memorial plaque commemorating Robert Watson-Watt and Arnold Wilkins

Development of radar greatly expanded on 1 September 1936 when Watson-Watt became Superintendent of a new establishment under the British Air Ministry, Bawdsey Research Station located in Bawdsey Manor, near Felixstowe, Suffolk. Work there resulted in the design and installation of aircraft detection and tracking stations called “Chain Home” along the East and South coasts of England in time for the outbreak of World War II in 1939. This system provided the vital advance information that helped the Royal Air Force win the Battle of Britain: without it, significant numbers of fighter aircraft would need to be always in the air to respond in short enough time if enemy aircraft detection relied on being physically seen by observers on the ground. Also vital was the “Dowding system” of reporting and coordination that was shown to be needed to make the best use of the radar information during tests of early deployment of radar in 1936 and 1937.

Given all required funding and development support, the team had working radar systems in 1935 and began deployment. By 1936, the first five Chain Home (CH) systems were operational and by 1940 stretched across the entire UK including Northern Ireland. Even by standards of the era, CH was crude; instead of broadcasting and receiving from an aimed antenna, CH broadcast a signal floodlighting the entire area in front of it, and then used one of Watt’s own radio direction finders to determine the direction of the returned echoes. This meant that CH transmitters had to be much more powerful and have better antennas than competing systems but allowed its rapid introduction using existing technologies.

During World War II

In April 1940, *Popular Science* showed an example of a radar unit using the Watson-Watt patent in an article on air defence. Also, in late 1941 *Popular Mechanics* had an article in which a U.S. scientist speculated about the British early warning system on the English east coast and came close to what it was and how it worked. Watson-Watt was sent to the U.S. in 1941 to advise on air defense after Japan’s attack on Pearl Harbor. Alfred Lee Loomis organized the Radiation Laboratory at Cambridge, Massachusetts which developed the technology in the years 1941–45. Later, in 1943, Page greatly improved radar with the monopulse technique that was used for many years in most radar applications.

The war precipitated research to find better resolution, more portability, and more features for radar, including complementary navigation systems like Oboe used by the RAF’s Pathfinder.

Applications

Commercial marine radar antenna. The rotating antenna radiates a vertical fan-shaped beam.

The information provided by radar includes the bearing and range (and therefore position) of the object from the radar scanner. It is thus used in many different fields where the need for such positioning is crucial. The first use of radar was for military purposes: to locate air, ground and sea targets. This evolved in the civilian field into applications for aircraft, ships, and roads.

In aviation, aircraft are equipped with radar devices that warn of aircraft or other obstacles in or approaching their path, display weather information, and give accurate altitude readings. The first commercial device fitted to aircraft was a 1938 Bell Lab unit on some United Air Lines aircraft. Such aircraft can land in fog at airports equipped with radar-assisted ground-controlled approach systems in which the plane's flight is observed on radar screens while operators radio landing directions to the pilot.

Marine radars are used to measure the bearing and distance of ships to prevent collision with other ships, to navigate, and to fix their position at sea when within range of shore or other fixed references such as islands, buoys, and lightships. In port or in harbour, vessel traffic service radar systems are used to monitor and regulate ship movements in busy waters.

Meteorologists use radar to monitor precipitation and wind. It has become the primary tool for short-term weather forecasting and watching for severe weather such as thunderstorms, tornadoes, winter storms, precipitation types, etc. Geologists use specialized ground-penetrating radars to map the composition of Earth's crust. Police forces use radar guns to monitor vehicle speeds on the roads. Smaller radar systems are used to detect human movement. Examples are breathing pattern detection for sleep monitoring and hand and finger gesture detection for computer interaction. Automatic door opening, light activation and intruder sensing are also common.

Principles

Radar Signal

A radar system has a transmitter that emits radio waves called *radar signals* in predetermined directions. When these come into contact with an object they are usually reflected or scattered in many directions. Radar signals are reflected especially well by materials of considerable electrical conductivity—especially by most metals, by seawater and by wet ground. Some of these make the use of radar altimeters possible. The radar signals that are reflected back towards the transmitter are the desirable ones that make radar work. If the object is *moving* either toward or away from the transmitter, there is a slight equivalent change in the frequency of the radio waves, caused by the Doppler effect.

Radar receivers are usually, but not always, in the same location as the transmitter. Although the reflected radar signals captured by the receiving antenna are usually very weak, they can be strengthened by electronic amplifiers. More sophisticated methods of signal processing are also used in order to recover useful radar signals.

The weak absorption of radio waves by the medium through which it passes is what enables radar sets to detect objects at relatively long ranges—ranges at which other electromagnetic wavelengths, such as visible light, infrared light, and ultraviolet light, are too strongly attenuated. Such weather phenomena as fog, clouds, rain, falling snow, and sleet that block visible light are usually transparent to radio waves. Certain radio frequencies that are absorbed or scattered by water vapour, raindrops, or atmospheric gases (especially oxygen) are avoided in designing radars, except when their detection is intended.

Illumination

Radar relies on its own transmissions rather than light from the Sun or the Moon, or from electromagnetic waves emitted by the objects themselves, such as infrared wavelengths (heat). This process of directing artificial radio waves towards objects is called *illumination*, although radio waves are invisible to the human eye or optical cameras.

Reflection

Brightness can indicate reflectivity as in this 1960 weather radar image (of Hurricane Abby).
The radar's frequency, pulse form, polarization, signal processing, and antenna determine what it can observe.

If electromagnetic waves travelling through one material meet another material, having a different dielectric constant or diamagnetic constant from the first, the waves will reflect or scatter from the boundary between the materials. This means that a solid object in air or in a vacuum, or a significant change in atomic density between the object and what is surrounding it, will usually scatter radar (radio) waves from its surface. This is particularly true for electrically conductive materials such as metal and carbon fibre, making radar well-suited to the detection of aircraft and ships. Radar absorbing material, containing resistive and sometimes magnetic substances, is used on military vehicles to reduce radar reflection. This is the radio equivalent of painting something a dark colour so that it cannot be seen by the eye at night.

Radar waves scatter in a variety of ways depending on the size (wavelength) of the radio wave and the shape of the target. If the wavelength is much shorter than the target's size, the wave will bounce off in a way similar to the way light is reflected by a mirror. If the wavelength is much longer than the size of the target, the target may not be visible because of poor reflection. Low-frequency radar technology is dependent on resonances for detection, but not identification, of targets. This is described by Rayleigh scattering, an effect that creates Earth's blue sky and red sunsets. When the two length scales are comparable, there may be resonances. Early radars used very long wavelengths that were larger than the targets and thus received a vague signal, where as some modern systems use shorter wavelengths (a few centimetres or less) that can image objects as small as a loaf of bread.

Short radio waves reflect from curves and corners in a way similar to glint from a rounded piece of glass. The most reflective targets for short wavelengths have 90° angles between the reflective surfaces. A corner reflector consists of three flat surfaces meeting like the inside corner of a box. The structure will reflect waves entering its opening directly back to the source. They are commonly used as radar reflectors to make otherwise difficult-to-detect objects easier to detect. Corner reflectors on boats, for example, make them more detectable to avoid collision or during a rescue. For similar reasons, objects intended to avoid detection will not have inside corners or surfaces and edges perpendicular to likely detection directions, which leads to "odd" looking stealth aircraft. These precautions do not completely eliminate reflection because of diffraction, especially at longer wavelengths. Half wavelength long wires or strips of conducting material, such as chaff, are very reflective but do not direct the scattered energy back toward the source. The extent to which an object reflects or scatters radio waves is called its radar cross section.

Radar Equation

The power P_r returning to the receiving antenna is given by the equation:

$$P_r = \frac{P_t G_t A_r \sigma F^4}{(4\pi)^2 R_t^2 R_r^2}$$

where

- P_t = transmitter power
- G_t = gain of the transmitting antenna
- A_r = effective aperture (area) of the receiving antenna; this can also be expressed as $\frac{G_r \lambda^2}{4\pi}$, where
    - λ = transmitted wavelength
    - G_r = gain of receiving antenna
- σ = radar cross section, or scattering coefficient, of the target
- F = pattern propagation factor
- R_t = distance from the transmitter to the target

- R_r = distance from the target to the receiver.

In the common case where the transmitter and the receiver are at the same location, $R_t = R_r$ and the term $R_t^2 R_r^2$ can be replaced by R^4, where R is the range. This yields:

$$P_r = \frac{P_t G_t A_r \sigma F^4}{(4\pi)^2 R^4}.$$

This shows that the received power declines as the fourth power of the range, which means that the received power from distant targets is relatively very small.

Additional filtering and pulse integration modifies the radar equation slightly for pulse-Doppler radar performance, which can be used to increase detection range and reduce transmit power.

The equation above with $F = 1$ is a simplification for transmission in a vacuum without interference. The propagation factor accounts for the effects of multipath and shadowing and depends on the details of the environment. In a real-world situation, pathloss effects should also be considered.

Doppler Effect

Frequency shift is caused by motion that changes the number of wavelengths between the reflector and the radar. That can degrade or enhance radar performance depending upon how that affects the detection process. As an example, Moving Target Indication can interact with Doppler to produce signal cancellation at certain radial velocities, which degrades performance.

Sea-based radar systems, semi-active radar homing, active radar homing, weather radar, military aircraft, and radar astronomy rely on the Doppler effect to enhance performance. This produces information about target velocity during the detection process. This also allows small objects to be detected in an environment containing much larger nearby slow moving objects.

Doppler shift depends upon whether the radar configuration is active or passive. Active radar transmits a signal that is reflected back to the receiver. Passive radar depends upon the object sending a signal to the receiver.

The Doppler frequency shift for active radar is as follows, where F_D is Doppler frequency, F_T is transmit frequency, V_R is radial velocity, and C is the speed of light:

$$F_D = 2 \times F_T \times \left(\frac{V_R}{C}\right).$$

Passive radar is applicable to electronic countermeasures and radio astronomy as follows:

$$F_D = F_T \times \left(\frac{V_R}{C}\right).$$

Only the radial component of the velocity is relevant. When the reflector is moving at right angle to the radar beam, it has no relative velocity. Vehicles and weather moving parallel to the radar beam produce the maximum Doppler frequency shift.

When the transmit frequency (F_T) is pulsed, using a pulse repeat frequency of F_R, the resulting frequency spectrum will contain harmonic frequencies above and below F_T with a distance of F_R. As a result, the Doppler measurement is only non-ambiguous if the Doppler frequency shift is less than half of F_R, called the Nyquist frequency, since the returned frequency otherwise cannot be distinguished from shifting of a harmonic frequency above or below, thus requiring:

$$|F_D| < \frac{F_R}{2}$$

Or when substituting with F_D:

$$|V_R| < \frac{F_R \times \frac{C}{F_T}}{4}$$

As an example, a Doppler weather radar with a pulse rate of 2 kHz and transmit frequency of 1 GHz can reliably measure weather speed up to at most 150 m/s (340 mph), thus cannot reliably determine radial velocity of aircraft moving 1,000 m/s (2,200 mph).

Polarization

In all electromagnetic radiation, the electric field is perpendicular to the direction of propagation, and the electric field direction is the polarization of the wave. For a transmitted radar signal, the polarization can be controlled to yield different effects. Radars use horizontal, vertical, linear, and circular polarization to detect different types of reflections. For example, circular polarization is used to minimize the interference caused by rain. Linear polarization returns usually indicate metal surfaces. Random polarization returns usually indicate a fractal surface, such as rocks or soil, and are used by navigation radars.

Limiting Factors

Beam Path and Range

BEAM HEIGHT WITH DISTANCE (AGL)

$$H = \left(\sqrt{r^2 + (k_e a_e)^2 + 2 r k_e a_e \sin(\theta_e)}\right) - k_e a_e + h_e$$

r: distance $\quad$ k_e: 4/3 (standard refraction coefficient)

a_e: *Earth radius* $\quad$ θ_e: *Elevation angle*

h_e: *Height of radar above ground*

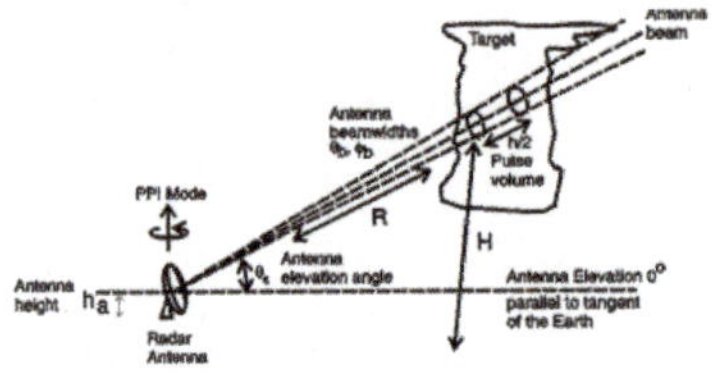

Echo heights above ground

The radar beam would follow a linear path in vacuum, but it really follows a somewhat curved path in the atmosphere because of the variation of the refractive index of air, that is called the radar horizon. Even when the beam is emitted parallel to the ground, it will rise above it as the Earth curvature sinks below the horizon. Furthermore, the signal is attenuated by the medium it crosses, and the beam disperses.

The maximum range of a conventional radar can be limited by a number of factors:

- Line of sight, which depends on height above ground. This means without a direct line of sight the path of the beam is blocked.
- The maximum non-ambiguous range, which is determined by the pulse repetition frequency. The maximum non-ambiguous range is the distance the pulse could travel and return before the next pulse is emitted.
- Radar sensitivity and power of the return signal as computed in the radar equation. This includes factors such as environmental conditions and the size (or radar cross section) of the target.

Noise

Signal noise is an internal source of random variations in the signal, which is generated by all electronic components.

Reflected signals decline rapidly as distance increases, so noise introduces a radar range limitation. The noise floor and signal to noise ratio are two different measures of performance that affect range performance. Reflectors that are too far away produce too little signal to exceed the noise floor and cannot be detected. Detection requires a signal that exceeds the noise floor by at least the signal to noise ratio.

Noise typically appears as random variations superimposed on the desired echo signal received in the radar receiver. The lower the power of the desired signal, the more difficult it is to discern it from the noise. Noise figure is a measure of the noise produced by a receiver compared to an ideal receiver, and this needs to be minimized.

Shot noise is produced by electrons in transit across a discontinuity, which occurs in all detectors. Shot noise is the dominant source in most receivers. There will also be flicker noise caused by electron transit through amplification devices, which is reduced using heterodyne amplification. Another reason for heterodyne processing is that for fixed fractional bandwidth, the instantaneous bandwidth increases linearly in frequency. This allows improved range resolution. The one notable exception to heterodyne (downconversion) radar systems is ultra-wideband radar. Here a single cycle, or transient wave, is used similar to UWB communications.

Noise is also generated by external sources, most importantly the natural thermal radiation of the background surrounding the target of interest. In modern radar systems, the internal noise is typically about equal to or lower than the external noise. An exception is if the radar is aimed upwards at clear sky, where the scene is so "cold" that it generates very little thermal noise. The thermal noise is given by $k_B\,T\,B$, where T is temperature, B is bandwidth (post matched filter) and k_B is Boltzmann's constant. There is an appealing intuitive interpretation of this relationship in a radar. Matched filtering allows the entire energy received from a target to be compressed into a single bin (be it a range, Doppler, elevation, or azimuth bin). On the surface it would appear that then within a fixed interval of time one could obtain perfect, error free, detection. To do this one simply compresses all energy into an infinitesimal time slice. What limits this approach in the real world is that, while time is arbitrarily divisible, current is not. The quantum of electrical energy is an electron, and so the best one can do is match filter all energy into a single electron. Since the electron is moving at a certain temperature (Plank spectrum) this noise source cannot be further eroded. We see then that radar, like all macro-scale entities, is profoundly impacted by quantum theory.

Noise is random and target signals are not. Signal processing can take advantage of this phenomenon to reduce the noise floor using two strategies. The kind of signal integration used with moving target indication can improve noise up to $\sqrt{2}$ for each stage. The signal can also be split among multiple filters for pulse-Doppler signal processing, which reduces the noise floor by the number of filters. These improvements depend upon coherence.

Interference

Radar systems must overcome unwanted signals in order to focus on the targets of interest. These unwanted signals may originate from internal and external sources, both passive and active. The ability of the radar system to overcome these unwanted signals defines its signal-to-noise ratio (SNR). SNR is defined as the ratio of the signal power to the noise power within the desired signal; it compares the level of a desired target signal to the level of background noise (atmospheric noise and noise generated within the receiver). The higher a system's SNR the better it is at discriminating actual targets from noise signals.

Clutter

Clutter refers to radio frequency (RF) echoes returned from targets which are uninteresting to the radar operators. Such targets include natural objects such as ground, sea, and when not being tasked for meteorological purposes, precipitation (such as rain, snow or hail), sand storms, animals (especially birds), atmospheric turbulence, and other atmospheric effects, such as ionosphere reflections, meteor trails, and Hail spike. Clutter may also be returned from man-made objects such as buildings and, intentionally, by radar countermeasures such as chaff.

Some clutter may also be caused by a long radar waveguide between the radar transceiver and the antenna. In a typical plan position indicator (PPI) radar with a rotating antenna, this will usually be seen as a "sun" or "sunburst" in the centre of the display as the receiver responds to echoes from dust particles and misguided RF in the waveguide. Adjusting the timing between when the transmitter sends a pulse and when the receiver stage is enabled will generally reduce the sunburst without affecting the accuracy of the range, since most sunburst is caused by a diffused transmit pulse reflected before it leaves the antenna. Clutter is considered a passive interference source, since it only appears in response to radar signals sent by the radar.

Clutter is detected and neutralized in several ways. Clutter tends to appear static between radar scans; on subsequent scan echoes, desirable targets will appear to move, and all stationary echoes can be eliminated. Sea clutter can be reduced by using horizontal polarization, while rain is reduced with circular polarization (note that meteorological radars wish for the opposite effect, and therefore use linear polarization to detect precipitation). Other methods attempt to increase the signal-to-clutter ratio.

Clutter moves with the wind or is stationary. Two common strategies to improve measure or performance in a clutter environment are:

- Moving target indication, which integrates successive pulses and
- Doppler processing, which uses filters to separate clutter from desirable signals.

The most effective clutter reduction technique is pulse-Doppler radar. Doppler separates clutter from aircraft and spacecraft using a frequency spectrum, so individual signals can be separated from multiple reflectors located in the same volume using velocity differences. This requires a coherent transmitter. Another technique uses a moving target indicator that subtracts the receive signal from two successive pulses using phase to reduce signals from slow moving objects. This can be adapted for systems that lack a coherent transmitter, such as time-domain pulse-amplitude radar.

Constant false alarm rate, a form of automatic gain control (AGC), is a method that relies on clutter returns far outnumbering echoes from targets of interest. The receiver's gain is automatically adjusted to maintain a constant level of overall visible clutter. While this does not help detect targets masked by stronger surrounding clutter, it does help to distinguish strong target sources. In the past, radar AGC was electronically controlled and affected the gain of the entire radar receiver. As radars evolved, AGC became computer-software controlled and affected the gain with greater granularity in specific detection cells.

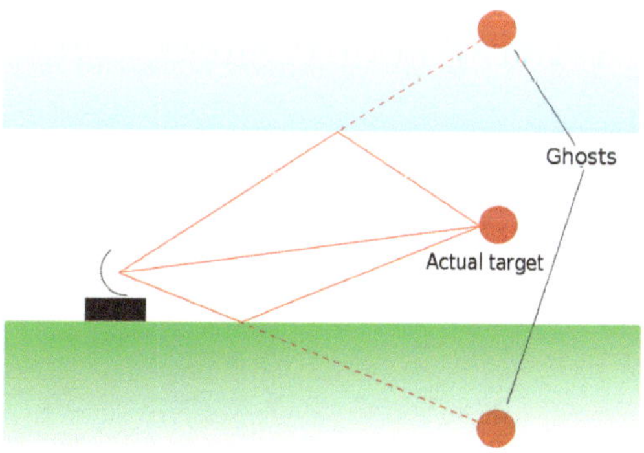

Radar multipath echoes from a target cause ghosts to appear.

Clutter may also originate from multipath echoes from valid targets caused by ground reflection, atmospheric ducting or ionospheric reflection/refraction (e.g., anomalous propagation). This clutter type is especially bothersome since it appears to move and behave like other normal (point) targets of interest. In a typical scenario, an aircraft echo is reflected from the ground below, appearing to the receiver as an identical target below the correct one. The radar may try to unify the targets, reporting the target at an incorrect height, or eliminating it on the basis of jitter or a physical impossibility. Terrain bounce jamming exploits this response by amplifying the radar signal and directing it downward. These problems can be overcome by incorporating a ground map of the radar's surroundings and eliminating all echoes which appear to originate below ground or above a certain height. Monopulse can be improved by altering the elevation algorithm used at low elevation. In newer air traffic control radar equipment, algorithms are used to identify the false targets by comparing the current pulse returns to those adjacent, as well as calculating return improbabilities.

Jamming

Radar jamming refers to radio frequency signals originating from sources outside the radar, transmitting in the radar's frequency and thereby masking targets of interest. Jamming may be intentional, as with an electronic warfare tactic, or unintentional, as with friendly forces operating equipment that transmits using the same frequency range. Jamming is considered an active interference source, since it is initiated by elements outside the radar and in general unrelated to the radar signals.

Jamming is problematic to radar since the jamming signal only needs to travel one way (from the jammer to the radar receiver) whereas the radar echoes travel two ways (radar-target-radar) and are therefore significantly reduced in power by the time they return to the radar receiver. Jammers therefore can be much less powerful than their jammed radars and still effectively mask targets along the line of sight from the jammer to the radar (*mainlobe jamming*). Jammers have an added effect of affecting radars along other lines of sight through the radar receiver's sidelobes (*sidelobe jamming*).

Mainlobe jamming can generally only be reduced by narrowing the mainlobe solid angle and cannot fully be eliminated when directly facing a jammer which uses the same frequency and polarization as the radar. Sidelobe jamming can be overcome by reducing receiving sidelobes in the radar antenna design and by using an omnidirectional antenna to detect and disregard non-mainlobe signals. Other anti-jamming techniques are frequency hopping and polarization.

Radar Signal Processing

Transit Time

Pulse radar: The round-trip time for the radar pulse to get to the target and return is measured. The distance is proportional to this time.

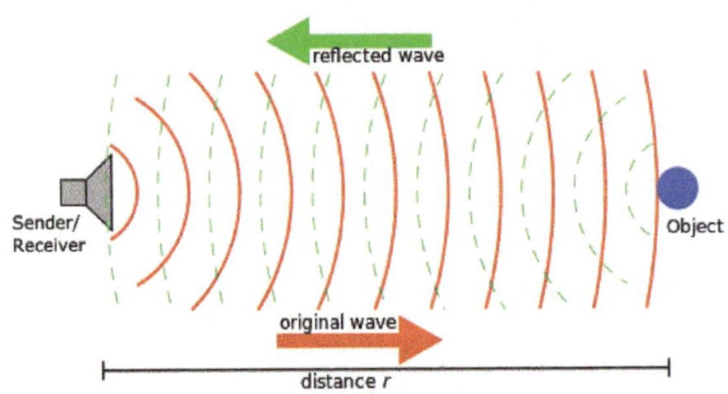

Continuous wave (CW) radar

One way to obtain a distance measurement is based on the time-of-flight: transmit a short pulse of radio signal (electromagnetic radiation) and measure the time it takes for the reflection to return. The distance is one-half the product of the round trip time (because the signal has to travel to the target and then back to the receiver) and the speed of the signal. Since radio waves travel at the speed of light, accurate distance measurement requires high-speed electronics. In most cases, the receiver does not detect the return while the signal is being transmitted. Through the use of a duplexer, the radar switches between transmitting and receiving at a predetermined rate. A similar effect imposes a maximum range as well. In order to maximize range, longer times between pulses should be used, referred to as a pulse repetition time, or its reciprocal, pulse repetition frequency.

These two effects tend to be at odds with each other, and it is not easy to combine both good short range and good long range in a single radar. This is because the short pulses needed for a good minimum range broadcast have less total energy, making the returns much smaller and the target harder to detect. This could be offset by using more pulses, but this would shorten the maximum range. So each radar uses a particular type of signal. Long-range radars tend to use long pulses with long delays between them, and short range radars use smaller pulses with less time between them. As electronics have improved many radars now can change their pulse repetition frequency, thereby changing their range. The newest radars fire two pulses during one cell, one for short range (about 10 km (6.2 mi)) and a separate signal for longer ranges (about 100 km (62 mi)).

The distance resolution and the characteristics of the received signal as compared to noise depends on the shape of the pulse. The pulse is often modulated to achieve better performance using a technique known as pulse compression.

Distance may also be measured as a function of time. The radar mile is the amount of time it takes for a radar pulse to travel one nautical mile, reflect off a target, and return to the radar antenna. Since a nautical mile is defined as 1,852 m, then dividing this distance by the speed of light (299,792,458 m/s), and then multiplying the result by 2 yields a result of 12.36 μs in duration.

Frequency Modulation

Another form of distance measuring radar is based on frequency modulation. Frequency comparison between two signals is considerably more accurate, even with older electronics, than timing the signal. By measuring the frequency of the returned signal and comparing that with the original, the difference can be easily measured.

This technique can be used in continuous wave radar and is often found in aircraft radar altimeters. In these systems a "carrier" radar signal is frequency modulated in a predictable way, typically varying up and down with a sine wave or sawtooth pattern at audio frequencies. The signal is then sent out from one antenna and received on another, typically located on the bottom of the aircraft, and the signal can be continuously compared using a simple *beat frequency* modulator that produces an audio frequency tone from the returned signal and a portion of the transmitted signal.

Since the signal frequency is changing, by the time the signal returns to the aircraft the transmit frequency has changed. The amount of frequency shift is used to measure distance.

The modulation index riding on the receive signal is proportional to the time delay between the radar and the reflector. The amount of that frequency shift becomes greater with greater time delay. The measure of the amount of frequency shift is directly proportional to the distance travelled. That distance can be displayed on an instrument, and it may also be available via the transponder. This signal processing is similar to that used in speed detecting Doppler radar. Example systems using this approach are AZUSA, MISTRAM, and UDOP.

A further advantage is that the radar can operate effectively at relatively low frequencies. This was important in the early development of this type when high frequency signal generation was difficult or expensive.

Terrestrial radar uses low-power FM signals that cover a larger frequency range. The multiple reflections are analyzed mathematically for pattern changes with multiple passes creating a computerized synthetic image. Doppler effects are used which allows slow moving objects to be detected as well as largely eliminating "noise" from the surfaces of bodies of water.

Speed Measurement

Speed is the change in distance to an object with respect to time. Thus the existing system for measuring distance, combined with a memory capacity to see where the target last was, is enough to measure speed. At one time the memory consisted of a user making grease pencil marks on the radar screen and then calculating the speed using a slide rule. Modern radar systems perform the equivalent operation faster and more accurately using computers.

If the transmitter's output is coherent (phase synchronized), there is another effect that can be used to make almost instant speed measurements (no memory is required), known as the Doppler effect. Most modern radar systems use this principle into Doppler radar and pulse-Doppler radar systems (weather radar, military radar). The Doppler effect is only able to determine the relative speed of the target along the line of sight from the radar to the target. Any component of target velocity perpendicular to the line of sight cannot be determined by using the Doppler effect alone, but it can be determined by tracking the target's azimuth over time.

It is possible to make a Doppler radar without any pulsing, known as a continuous-wave radar (CW radar), by sending out a very pure signal of a known frequency. CW radar is ideal for determining the radial component of a target's velocity. CW radar is typically used by traffic enforcement to measure vehicle speed quickly and accurately where range is not important.

When using a pulsed radar, the variation between the phase of successive returns gives the distance the target has moved between pulses, and thus its speed can be calculated. Other mathematical developments in radar signal processing include time-frequency analysis (Weyl Heisenberg or wavelet), as well as the chirplet transform which makes use of the change of frequency of returns from moving targets ("chirp").

Pulse-Doppler Signal Processing

Pulse-Doppler signal processing includes frequency filtering in the detection process. The space between each transmit pulse is divided into range cells or range gates. Each cell is filtered independently much like the process used by a spectrum analyzer to produce the display showing different frequencies. Each different distance produces a different spectrum. These spectra are used to perform the detection process. This is required to achieve acceptable performance in hostile environments involving weather, terrain, and electronic countermeasures.

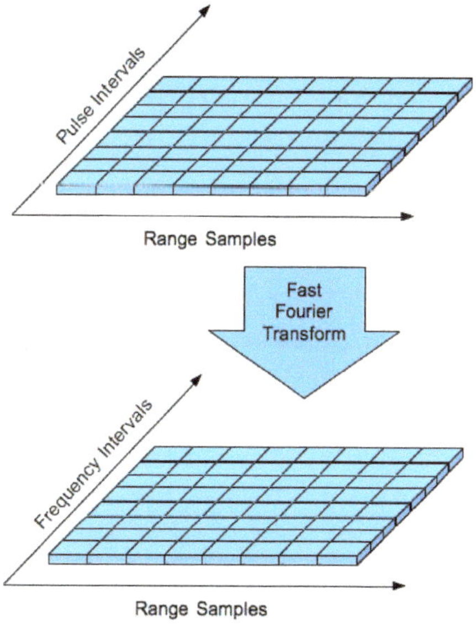

Pulse-Doppler signal processing. The *Range Sample* axis represents individual samples taken in between each transmit pulse. The *Range Interval* axis represents each successive transmit pulse interval during which samples are taken. The Fast Fourier Transform process converts time-domain samples into frequency domain spectra. This is sometimes called the *bed of nails*.

The primary purpose is to measure both the amplitude and frequency of the aggregate reflected signal from multiple distances. This is used with weather radar to measure radial wind velocity and precipitation rate in each different volume of air. This is linked with computing systems to produce a real-time electronic weather map. Aircraft safety depends upon continuous access to accurate weather radar information that is used to prevent injuries and accidents. Weather radar uses a low PRF. Coherency requirements are not as strict as those for military systems because individual signals ordinarily do not need to be separated. Less sophisticated filtering is required, and range ambiguity processing is not normally needed with weather radar in comparison with military radar intended to track air vehicles.

The alternate purpose is "look-down/shoot-down" capability required to improve military air combat survivability. Pulse-Doppler is also used for ground based surveillance radar required to defend personnel and vehicles. Pulse-Doppler signal processing increases the maximum detection distance using less radiation in close proximity to aircraft pilots, shipboard personnel, infantry, and artillery. Reflections from terrain, water, and weather produce signals much larger than aircraft and missiles, which allows fast moving vehicles to hide using nap-of-the-earth flying techniques and stealth technology to avoid detection until an attack vehicle is too close to destroy. Pulse-Doppler signal processing incorporates more sophisticated electronic filtering that safely eliminates this kind of weakness. This requires the use of medium pulse-repetition frequency with phase coherent hardware that has a large dynamic range. Military applications require medium PRF which prevents range from being determined directly, and range ambiguity resolution processing is required to identify the true range of all reflected signals. Radial movement is usually linked with Doppler frequency to produce a lock signal that cannot be produced by radar jamming signals. Pulse-Doppler signal processing also produces audible signals that can be used for threat identification.

Reduction of Interference Effects

Signal processing is employed in radar systems to reduce the radar interference effects. Signal processing techniques include moving target indication, Pulse-Doppler signal processing, moving target detection processors, correlation with secondary surveillance radar targets, space-time adaptive processing, and track-before-detect. Constant false alarm rate and digital terrain model processing are also used in clutter environments.

Plot and Track Extraction

A Track algorithm is a radar performance enhancement strategy. Tracking algorithms provide the ability to predict future position of multiple moving objects based on the history of the individual positions being reported by sensor systems.

Historical information is accumulated and used to predict future position for use with air traffic control, threat estimation, combat system doctrine, gun aiming, and missile guidance. Position data is accumulated radar sensors over the span of a few minutes.

There are four common track algorithms.

- Nearest neighbour algorithm
- Probabilistic Data Association
- Multiple Hypothesis Tracking
- Interactive Multiple Model (IMM)

Radar video returns from aircraft can be subjected to a plot extraction process whereby spurious and interfering signals are discarded. A sequence of target returns can be monitored through a device known as a plot extractor.

The non-relevant real time returns can be removed from the displayed information and a single plot displayed. In some radar systems, or alternatively in the command and control system to which the radar is connected, a radar tracker is used to associate the sequence of plots belonging to individual targets and estimate the targets' headings and speeds.

Engineering

A radar's components are:

- A transmitter that generates the radio signal with an oscillator such as a klystron or a magnetron and controls its duration by a modulator.
- A waveguide that links the transmitter and the antenna.
- A duplexer that serves as a switch between the antenna and the transmitter or the receiver for the signal when the antenna is used in both situations.
- A receiver. Knowing the shape of the desired received signal (a pulse), an optimal receiver can be designed using a matched filter.

- A display processor to produce signals for human readable output devices.
- An electronic section that controls all those devices and the antenna to perform the radar scan ordered by software.
- A link to end user devices and displays.

Components of a Radar/Composantes d'un radar

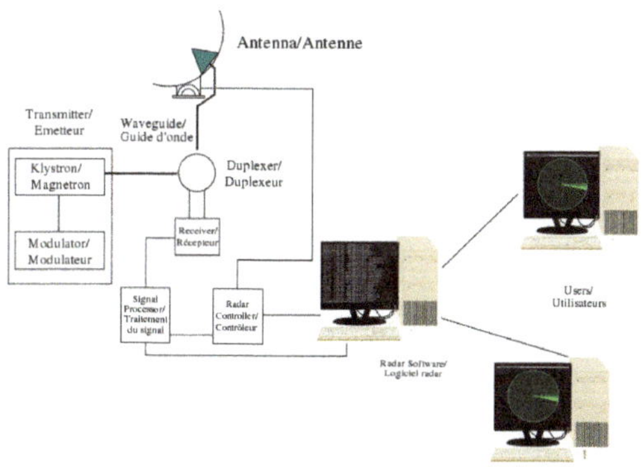

Radar components

Antenna Design

Radio signals broadcast from a single antenna will spread out in all directions, and likewise a single antenna will receive signals equally from all directions. This leaves the radar with the problem of deciding where the target object is located.

Early systems tended to use omnidirectional broadcast antennas, with directional receiver antennas which were pointed in various directions. For instance, the first system to be deployed, Chain Home, used two straight antennas at right angles for reception, each on a different display. The maximum return would be detected with an antenna at right angles to the target, and a minimum with the antenna pointed directly at it (end on). The operator could determine the direction to a target by rotating the antenna so one display showed a maximum while the other showed a minimum. One serious limitation with this type of solution is that the broadcast is sent out in all directions, so the amount of energy in the direction being examined is a small part of that transmitted. To get a reasonable amount of power on the "target", the transmitting aerial should also be directional.

Parabolic Reflector

More modern systems use a steerable parabolic "dish" to create a tight broadcast beam, typically using the same dish as the receiver. Such systems often combine two radar frequencies in the same antenna in order to allow automatic steering, or *radar lock*.

Parabolic reflectors can be either symmetric parabolas or spoiled parabolas: Symmetric parabolic antennas produce a narrow "pencil" beam in both the X and Y dimensions and consequently have a higher gain. The NEXRAD Pulse-Doppler weather radar uses a symmetric antenna to perform detailed volumetric scans of the atmosphere. Spoiled parabolic antennas produce a narrow beam in one dimension and a relatively wide beam in the other. This feature is useful if target detection over a wide range of angles is more important than target location in three dimensions. Most 2D surveillance radars use a spoiled parabolic antenna with a narrow azimuthal beamwidth and wide vertical beamwidth. This beam configuration allows the radar operator to detect an aircraft at a specific azimuth but at an indeterminate height. Conversely, so-called "nodder" height finding radars use a dish with a narrow vertical beamwidth and wide azimuthal beamwidth to detect an aircraft at a specific height but with low azimuthal precision.

Surveillance radar antenna

Types of Scan

- Primary Scan: A scanning technique where the main antenna aerial is moved to produce a scanning beam, examples include circular scan, sector scan, etc.
- Secondary Scan: A scanning technique where the antenna feed is moved to produce a scanning beam, examples include conical scan, unidirectional sector scan, lobe switching, etc.
- Palmer Scan: A scanning technique that produces a scanning beam by moving the main antenna and its feed. A Palmer Scan is a combination of a Primary Scan and a Secondary Scan.
- Conical scanning: The radar beam is rotated in a small circle around the "boresight" axis, which is pointed at the target.

Slotted Waveguide

Slotted waveguide antenna

Applied similarly to the parabolic reflector, the slotted waveguide is moved mechanically to scan and is particularly suitable for non-tracking surface scan systems, where the vertical pattern may remain constant. Owing to its lower cost and less wind exposure, shipboard, airport surface, and harbour surveillance radars now use this approach in preference to a parabolic antenna.

Phased Array

Phased array: Not all radar antennas must rotate to scan the sky.

Another method of steering is used in a phased array radar.

Phased array antennas are composed of evenly spaced similar antenna elements, such as aerials or rows of slotted waveguide. Each antenna element or group of antenna elements incorporates a discrete phase shift that produces a phase gradient across the array. For example, array elements producing a 5 degree phase shift for each wavelength across the array face will produce a beam pointed 5 degree away from the centreline perpendicular to the array face. Signals travelling along that beam will be reinforced. Signals offset from that beam will be cancelled. The amount of reinforcement is antenna gain. The amount of cancellation is side-lobe suppression.

Phased array radars have been in use since the earliest years of radar in World War II (Mammut radar), but electronic device limitations led to poor performance. Phased array radars were originally used for missile defence. They are the heart of the ship-borne Aegis Combat System and the Patriot Missile System. The massive redundancy associated with having a large number of array elements increases reliability at the expense of gradual performance degradation that occurs as individual phase elements fail. To a lesser extent, Phased array radars have been used in Weather Surveillance. As of 2017, NOAA plans to implement a national network of Multi-Function Phased array radars throughout the United States within 10 years, for meteorological studies and flight monitoring.

Phased array antenna can be built to conform to specific shapes, like missiles, infantry support vehicles, ships, and aircraft.

As the price of electronics has fallen, phased array radars have become more common. Almost all modern military radar systems are based on phased arrays, where the small additional cost is offset by the improved reliability of a system with no moving parts. Traditional moving-antenna designs are still widely used in roles where cost is a significant factor such as air traffic surveillance and similar systems.

Phased array radars are valued for use in aircraft since they can track multiple targets. The first aircraft to use a phased array radar was the B-1B Lancer. The first fighter aircraft to use phased array radar was the Mikoyan MiG-31. The MiG-31M's SBI-16 Zaslon Passive electronically scanned array radar was considered to be the world's most powerful fighter radar, until the AN/APG-77 Active electronically scanned array was introduced on the Lockheed Martin F-22 Raptor.

Phased-array interferometry or aperture synthesis techniques, using an array of separate dishes that are phased into a single effective aperture, are not typical for radar applications, although they are widely used in radio astronomy. Because of the thinned array curse, such multiple aperture arrays, when used in transmitters, result in narrow beams at the expense of reducing the total power transmitted to the target. In principle, such techniques could increase spatial resolution, but the lower power means that this is generally not effective.

Aperture synthesis by post-processing motion data from a single moving source, on the other hand, is widely used in space and airborne radar systems.

Frequency Bands

The traditional band names originated as code-names during World War II and are still in military and aviation use throughout the world. They have been adopted in the United States by the Institute of Electrical and Electronics Engineers and internationally by the International Telecommunication Union. Most countries have additional regulations to control which parts of each band are available for civilian or military use.

Other users of the radio spectrum, such as the broadcasting and electronic countermeasures industries, have replaced the traditional military designations with their own systems.

Radar frequency bands			
Band name	**Frequency range**	**Wavelength range**	**Notes**
HF	3–30 MHz	10–100 m	Coastal radar systems, over-the-horizon radar (OTH) radars; 'high frequency'
VHF	30–300 MHz	1–10 m	Very long range, ground penetrating; 'very high frequency'
P	< 300 MHz	> 1 m	'P' for 'previous', applied retrospectively to early radar systems; essentially HF + VHF
UHF	300–1000 MHz	0.3–1 m	Very long range (e.g. ballistic missile early warning), ground penetrating, foliage penetrating; 'ultra high frequency'
L	1–2 GHz	15–30 cm	Long range air traffic control and surveillance; 'L' for 'long'
S	2–4 GHz	7.5–15 cm	Moderate range surveillance, Terminal air traffic control, long-range weather, marine radar; 'S' for 'short'
C	4–8 GHz	3.75–7.5 cm	Satellite transponders; a compromise (hence 'C') between X and S bands; weather; long range tracking
X	8–12 GHz	2.5–3.75 cm	Missile guidance, marine radar, weather, medium-resolution mapping and ground surveillance; in the United States the narrow range 10.525 GHz ±25 MHz is used for airport radar; short range tracking. Named X band because the frequency was a secret during WW2.
K_u	12–18 GHz	1.67–2.5 cm	High-resolution, also used for satellite transponders, frequency under K band (hence 'u')

K	18–24 GHz	1.11–1.67 cm	From German *kurz*, meaning 'short'; limited use due to absorption by water vapour, so K_u and K_a were used instead for surveillance. K-band is used for detecting clouds by meteorologists, and by police for detecting speeding motorists. K-band radar guns operate at 24.150 ± 0.100 GHz.
K_a	24–40 GHz	0.75–1.11 cm	Mapping, short range, airport surveillance; frequency just above K band (hence 'a') Photo radar, used to trigger cameras which take pictures of license plates of cars running red lights, operates at 34.300 ± 0.100 GHz.
mm	40–300 GHz	1.0–7.5 mm	Millimetre band, subdivided as below. The frequency ranges depend on waveguide size. Multiple letters are assigned to these bands by different groups. These are from Baytron, a now defunct company that made test equipment.
V	40–75 GHz	4.0–7.5 mm	Very strongly absorbed by atmospheric oxygen, which resonates at 60 GHz.
W	75–110 GHz	2.7–4.0 mm	Used as a visual sensor for experimental autonomous vehicles, high-resolution meteorological observation, and imaging.

Radar Modulators

Modulators act to provide the waveform of the RF-pulse. There are two different radar modulator designs:

- High voltage switch for non-coherent keyed power-oscillators These modulators consist of a high voltage pulse generator formed from a high voltage supply, a pulse forming network, and a high voltage switch such as a thyratron. They generate short pulses of power to feed, e.g., the magnetron, a special type of vacuum tube that converts DC (usually pulsed) into microwaves. This technology is known as pulsed power. In this way, the transmitted pulse of RF radiation is kept to a defined and usually very short duration.

- Hybrid mixers, fed by a waveform generator and an exciter for a complex but coherent waveform. This waveform can be generated by low power/low-voltage input signals. In this case the radar transmitter must be a power-amplifier, e.g., a klystron tube or a solid state transmitter. In this way, the transmitted pulse is intrapulse-modulated and the radar receiver must use pulse compression techniques.

Radar Coolant

Coherent microwave amplifiers operating above 1,000 watts microwave output, like travelling wave tubes and klystrons, require liquid coolant. The electron beam must contain 5 to 10 times more power than the microwave output, which can produce enough heat to generate plasma. This plasma flows from the collector toward the cathode. The same magnetic focusing that guides the electron beam forces the plasma into the path of the electron beam but flowing in the opposite direction. This introduces FM modulation which degrades Doppler performance. To prevent this, liquid coolant with minimum pressure and flow rate is required, and deionized water is normally used in most high power surface radar systems that utilize Doppler processing.

Coolanol (silicate ester) was used in several military radars in the 1970s. However, it is hygroscopic, leading to hydrolysis and formation of highly flammable alcohol. The loss of a U.S. Navy aircraft in 1978 was attributed to a silicate ester fire. Coolanol is also expensive and toxic. The U.S. Navy has instituted a program named Pollution Prevention (P2) to eliminate or reduce the volume and toxicity of waste, air emissions, and effluent discharges. Because of this, Coolanol is used less often today.

Regulations

Radar is defined by *article 1.100* of the International Telecommunication Union´s (ITU) ITU Radio Regulations (RR) as:

A radiodetermination system based on the comparison of reference signals with radio signals reflected, or retransmitted, from the position to be determined. Each *radiodetermination system* shall be classified by the *radiocommunication service* in which it operates permanently or temporarily. Typical radar utilizations are primary radar and secondary radar, these might operate in the radiolocation service or the radiolocation-satellite service.

History of Radar

The history of radar started with experiments by Heinrich Hertz in the late 19th century that showed that radio waves were reflected by metallic objects. This possibility was suggested in James Clerk Maxwell's seminal work on electromagnetism. However, it was not until the early 20th century that systems able to use these principles were becoming widely available, and it was German inventor Christian Hülsmeyer who first used them to build a simple ship detection device intended to help avoid collisions in fog (Reichspatent Nr. 165546). Numerous similar systems, which provided directional information to objects over short ranges, were developed over the next two decades.

A Chain Home transmitter antenna, part of one of the first comprehensive radar systems.

The development of systems able to produce short pulses of radio energy was the key advance that allowed modern radar systems to come into existence. By timing the pulses on an oscilloscope, the range could be determined and the direction of the antenna revealed the angular location of the targets. The two, combined, produced a "fix", locating the target relative to the antenna. In the 1934–1939 period, eight nations developed independently, and in great secrecy, systems of this type: the United Kingdom, Germany, the United States, the USSR, Japan, the Netherlands, France, and Italy. In addition, Britain shared their information with the United States and four Commonwealth countries: Australia, Canada, New Zealand, and South Africa, and these countries also developed their own radar systems. During the war, Hungary was added to this list. The term *RADAR* was coined in 1939 by the United States Signal Corps as it worked on these systems for the Navy.

The German *Freya* worked at higher frequencies, and was thus smaller than its Chain Home counterpart.

The anode block of the original cavity magnetron built by Randal and Boot, which provided a leap forward in radar design.

Progress during the war was rapid and of great importance, probably one of the decisive factors for the victory of the Allies. A key development was the magnetron in the UK, which allowed the creation of relatively small systems with sub-meter resolution. By the end of hostilities, Britain, Germany, the United States, the USSR, and Japan had a wide diversity of land- and sea-based radars as well as small airborne systems. After the war, radar use was widened to numerous fields including: civil aviation, marine navigation, radar guns for police, meteorology and even medicine. Key developments in the post-war period include the travelling wave tube as a way to produce large quantities of coherent microwaves, the development of signal delay systems that led to phased array radars, and ever-increasing frequencies that allow higher resolutions. Increases in signal processing capability due to the introduction of solid state computers has also had a large impact on radar use.

Significance

The place of radar in the larger story of science and technology is argued differently by different authors. On the one hand, radar contributed very little to theory, which was largely known since the days of Maxwell and Hertz. Therefore, radar did not advance science, but was simply a matter of technology and engineering. Maurice Ponte, one of the developers of radar in France, states:

The fundamental principle of the radar belongs to the common patrimony of the physicists; after all, what is left to the real credit of the technicians is measured by the effective realisation of operational materials.

But others point out the immense practical consequences of the development of radar. Far more than the atomic bomb, radar contributed to the Allied victory in World War II. Robert Buderi states that it was also the precursor of much modern technology. From a review of his book:

"... radar has been the root of a wide range of achievements since the war, producing a veritable family tree of modern technologies. Because of radar, astronomers can map the contours of far-off planets, physicians can see images of internal organs, meteorologists can measure rain falling in distant places, air travel is hundreds of times safer than travel by road, long-distance telephone calls are cheaper than postage, computers have become ubiquitous and ordinary people can cook their daily dinners in the time between sitcoms, with what used to be called a *radar range*".

In later years radar was used in scientific instruments, such as weather radar and radar astronomy.

Early Contributors

Heinrich Hertz

In 1886-1888 the German physicist Heinrich Hertz conducted his series of experiments that proved the existence of electromagnetic waves (including radio waves), predicted in equations developed in 1862–4 by the Scottish physicist James Clerk Maxwell. In Hertz's 1887 experiment he found that these waves would transmit through different types of materials and also would reflect off metal surfaces in his lab as well as conductors and dielectrics. The nature of these waves being similar to visible light in their ability to be reflected, refracted, and polarized would be shown by Hertz and subsequent experiments by other physicists.

Guglielmo Marconi

Radio pioneer Guglielmo Marconi noticed radio waves were being reflected back to the transmitter by objects in radio beacon experiments he conducted on March 3, 1899 on Salisbury Plain. In 1916 he and a British engineer called Charles Samuel Franklin used short-waves in their experiments, critical to the practical development of radar. He would relate his findings 6 years later in a 1922 paper delivered before the Institution of Electrical Engineers in London:

"I also described tests carried out in transmitting a beam of reflected waves across country ... and pointed out the possibility of the utility of such a system if applied to lighthouses and lightships, so as to enable vessels in foggy weather to locate dangerous points around the coasts ..."

It [now] seems to me that it should be possible to design [an] apparatus by means of which a ship could radiate or project a divergent beam of these rays in any desired direction, which rays, if coming across a metallic object, such as another steamer or ship, would be reflected back to a receiver screened from the local transmitter on the sending ship, and thereby immediately reveal the presence and bearing of the other ship in fog or thick weather.

Christian Hülsmeyer

In 1904, Christian Hülsmeyer gave public demonstrations in Germany and the Netherlands of the use of radio echoes to detect ships so that collisions could be avoided. His device consisted of a simple spark gap used to generate a signal that was aimed using a dipole antenna with a cylindrical parabolic reflector. When a signal reflected from a ship was picked up by a similar antenna attached to the separate coherer receiver, a bell sounded. During bad weather or fog, the device would be periodically spun to check for nearby ships. The apparatus detected the presence of ships up to 3 kilometres (1.6 nmi), and Hülsmeyer planned to extend its capability to 10 kilometres (5.4 nmi). It did not provide range (distance) information, only warning of a nearby object. He patented the device, called the *telemobiloscope*, but due to lack of interest by the naval authorities the invention was not put into production.

Hülsmeyer also received a patent amendment for estimating the range to the ship. Using a vertical scan of the horizon with the *telemobiloscope* mounted on a tower, the operator would find the angle at which the return was the most intense and deduce, by simple triangulation, the approximate distance. This is in contrast to the later development of pulsed radar, which determines distance via two-way transit time of the pulse.

United Kingdom

Robert Watson-Watt

In 1915, Robert Watson Watt joined the Meteorological Office as a meteorologist, working at an outstation at Aldershot in Hampshire. Over the next 20 years, he studied atmospheric phenomena and developed the use of radio signals generated by lightning strikes to map out the position of thunderstorms. The difficulty in pinpointing the direction of these fleeting signals using rotatable directional antennas led, in 1923, to the use of oscilloscopes in order to display the signals. The operation eventually moved to the outskirts of Slough in Berkshire, and in 1927 formed the Radio Research Station (RRS), Slough, an entity under the Department of Scientific and Industrial Research (DSIR). Watson Watt was appointed the RRS Superintendent.

As war clouds gathered over Britain, the likelihood of air raids and the threat of invasion by air and sea drove a major effort in applying science and technology to defence. In November 1934, the Air Ministry established the Committee for the Scientific Survey of Air Defence (CSSAD) with the official function of considering "how far recent advances in scientific and technical knowledge can be used to strengthen the present methods of defence against hostile aircraft". Commonly called the "Tizard Committee" after its Chairman, Sir Henry Tizard, this group had a profound influence on technical developments in Britain.

H. E. Wimperis, Director of Scientific Research at the Air Ministry and a member of the Tizard Committee, had read about a German newspaper article claiming that the Germans had built a death ray using radio signals, accompanied by an image of a very large radio antenna. Both concerned and potentially excited by this possibility, but highly skeptical at the same time, Wimperis looked for an expert in the field of radio propagation who might be able to pass judgement on the concept. Watt, Superintendent of the RRS, was now well established as an authority in the field of radio, and in January 1935, Wimperis contacted him asking if radio might be used for such a device. After discussing this with his scientific assistant, Arnold F. 'Skip' Wilkins, Wilkins quickly produced a back-of-the-envelope calculation that showed the energy required would be enormous. Watt wrote back that this was unlikely, but added the following comment: "Attention is being turned to the still difficult, but less unpromising, problem of radio detection and numerical considerations on the method of detection by reflected radio waves will be submitted when required".

Over the following several weeks, Wilkins considered the radio detection problem. He outlined an approach and backed it with detailed calculations of necessary transmitter power, reflection characteristics of an aircraft, and needed receiver sensitivity. He proposed using a directional receiver based on Watt's lightning detection concept, listening for powerful signals from a separate transmitter. Timing, and thus distance measurements, would be accomplished by triggering the oscilloscope's trace with a muted signal from the transmitter, and then simply measuring the returns against a scale. Watson Watt sent this information to the Air Ministry on February 12, 1935, in a secret report titled "The Detection of Aircraft by Radio Methods".

Reflection of radio signals was critical to the proposed technique, and the Air Ministry asked if this could be proven. To test this, Wilkins set up receiving equipment in a field near Upper Stowe, Northamptonshire. On February 26, 1935, a Handley Page Heyford bomber flew along a path between the receiving station and the transmitting towers of a BBC shortwave station in nearby Daventry. The aircraft reflected the 6 MHz (49 m) BBC signal, and this was readily detected by Arnold "Skip" Wilkins using Doppler-beat interference at ranges up to 8 mi (13 km). This convincing test, known as the *Daventry Experiment*, was witnessed by a representative from the Air Ministry, and led to the immediate authorization to build a full demonstration system. This experiment was later reproduced by Wilkins for the 1977 BBC television series *The Secret War* episode "To See a Hundred Miles".

Based on pulsed transmission as used for probing the ionosphere, a preliminary system was designed and built at the RRS by the team. Their existing transmitter had a peak power of about 1 kW, and Wilkins had estimated that 100 kW would be needed. Edward George Bowen was added to the team to design and build such a transmitter. Bowens' transmitter operated at 6 MHz (50 m), had a pulse-repetition rate of 25 Hz, a pulse width of 25 μs, and approached the desired power.

Orfordness, a narrow 19-mile (31 km) peninsula in Suffolk along the coast of the North Sea, was selected as the test site. Here the equipment would be openly operated in the guise of an ionospheric monitoring station. In mid-May 1935, the equipment was moved to Orfordness. Six wooden towers were erected, two for stringing the transmitting antenna, and four for corners of crossed receiving antennas. In June, general testing of the equipment began.

On June 17, the first target was detected—a Supermarine Scapa flying boat at 17 mi (27 km) range. It is historically correct that, on June 17, 1935, radio-based detection and ranging was first demonstrated in Britain. Watson Watt, Wilkins, and Bowen are generally credited with initiating what would later be called radar in this nation.

In December 1935, the British Treasury appropriated £60,000 for a five-station system called Chain Home (CH), covering approaches to the Thames Estuary. The secretary of the Tizard Committee, Albert Percival Rowe, coined the acronym RDF as a cover for the work, meaning Range and Direction Finding but suggesting the already well-known Radio Direction Finding.

Late in 1935, responding to Lindemann's recognition of the need for night detection and interception gear, and realizing existing transmitters were too heavy for aircraft, Bowen proposed fitting only receivers, what would later be called bistatic radar. Frederick Lindemann's proposals for infrared sensors and aerial mines would prove impractical. It would take Bowen's efforts, at the urging of Tizard, who became increasingly concerned about the need, to see Air to Surface Vessel (ASV), and through it Airborne Interception (AI), radar to fruition.

In 1937, Bowen's team set their crude ASV radar, the world's first airborne set, to detect the Home Fleet in dismal weather. Only in spring 1939, "as a matter of great urgency" after the failure of the searchlight system Silhouette, did attention turn to using ASV for air-to-air interception (AI). Demonstrated in June 1939, AI got a warm reception from Air Chief Marshal Hugh Dowding, and even more so from Churchill. This proved problematic. Its accuracy, dependent on the height of the aircraft, meant that CH, capable of only 4 sm (0.0068 km), was not accurate enough to place an aircraft within its detection range, and an additional system was required. Its wooden chassis had a disturbing tendency to catch fire (even with attention from expert technicians), so much so that Dowding, when told that Watson-Watt could provide hundreds of sets, demanded "ten that work". The Cossor and MetroVick sets were overweight for aircraft use and the RAF lacked night fighter pilots, observers, and suitable aircraft.

In 1940, John Randall and Harry Boot developed the cavity magnetron, which made ten-centimetre radar a reality. This device, the size of a small dinner plate, could be carried easily on aircraft and the short wavelength meant the antenna would also be small and hence suitable for mounting on aircraft. The short wavelength and high power made it very effective at spotting submarines from the air.

To aid Chain Home in making height calculations, at Dowding's request, the Electrical Calculator Type Q (commonly called the "Fruit Machine") was introduced in 1940.

The solution to night intercepts would be provided by Dr. W. B. “Ben” Lewis, who proposed a new, more accurate ground control display, the Plan Position Indicator (PPI), a new Ground-Controlled Interception (GCI) radar, and reliable AI radar. The AI sets would ultimately be built by EMI. GCI was unquestionably delayed by Watson-Watt’s opposition to it and his belief that CH was sufficient, as well as by Bowen’s preference for using ASV for navigation, despite Bomber Command disclaiming a need for it, and by Tizard’s reliance on the faulty Silhouette system.

Air Ministry

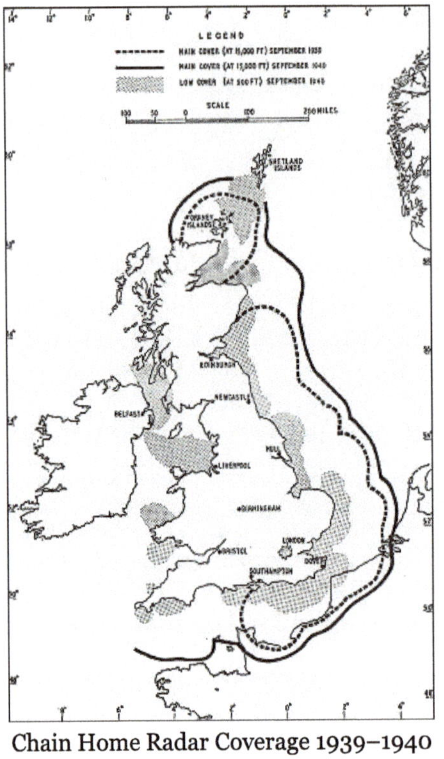

Chain Home Radar Coverage 1939–1940

In March 1936, the work at Orfordness was moved to Bawdsey Manor, nearby on the mainland. Until this time, the work had officially still been under the DSIR, but was now transferred to the Air Ministry. At the new Bawdsey Research Station, the Chain Home (CH) equipment was assembled as a prototype. There were equipment problems when the Royal Air Force (RAF) first exercised the prototype station in September 1936. These were cleared by the next April, and the Air Ministry started plans for a larger network of stations.

Initial hardware at CH stations was as follows: The transmitter operated on four pre-selected frequencies between 20 and 55 MHz, adjustable within 15 seconds, and delivered a peak power of 200 kW. The pulse duration was adjustable between 5 and 25 µs, with a repetition rate selectable as either 25 or 50 Hz. For synchronization of all CH transmitters, the pulse generator was locked to the 50 Hz of the British power grid. Four 360-foot (110 m) steel towers supported transmitting antennas, and four 240-foot (73 m) wooden towers supported cross-dipole arrays at three different levels. A goniometer was used to improve the directional accuracy from the multiple receiving antennas.

By the summer of 1937, 20 initial CH stations were in check-out operation. A major RAF exercise was performed before the end of the year, and was such a success that £10,000,000 was appropriated by the Treasury for an eventual full chain of coastal stations. At the start of 1938, the RAF took over control of all CH stations, and the network began regular operations.

In May 1938, Rowe replaced Watson Watt as Superintendent at Bawdsey. In addition to the work on CH and successor systems, there was now major work in airborne RDF equipment. This was led by E. G. Bowen and centered on 200-MHz (1.5 m) sets. The higher frequency allowed smaller antennas, appropriate for aircraft installation.

From the initiation of RDF work at Orfordness, the Air Ministry had kept the British Army and the Royal Navy generally informed; this led to both of these forces having their own RDF developments.

British Army

In 1931, at the Woolwich Research Station of the Army's Signals Experimental Establishment (SEE), W. A. S. Butement and P. E. Pollard had examined pulsed 600 MHz (50-cm) signals for detection of ships. Although they prepared a memorandum on this subject and performed preliminary experiments, for undefined reasons the War Office did not give it consideration.

As the Air Ministry's work on RDF progressed, Colonel Peter Worlledge of the Royal Engineer and Signals Board met with Watson Watt and was briefed on the RDF equipment and techniques being developed at Orfordness. His report, "The Proposed Method of Aeroplane Detection and Its Prospects", led the SEE to set up an "Army Cell" at Bawdsey in October 1936. This was under E. Talbot Paris and the staff included Butement and Pollard. The Cell's work emphasize two general types of RDF equipment: gun-laying (GL) systems for assisting anti-aircraft guns and searchlights, and coastal- defense (CD) systems for directing coastal artillery and defense of Army bases overseas.

Pollard led the first project, a gun-laying RDF code-named Mobile Radio Unit (MRU). This truck-mounted system was designed as a small version of a CH station. It operated at 23 MHz (13 m) with a power of 300 kW. A single 105-foot (32 m) tower supported a transmitting antenna, as well as two receiving antennas set orthogonally for estimating the signal bearing. In February 1937, a developmental unit detected an aircraft at a range of 60 miles (96 km). The Air Ministry also adopted this system as a mobile auxiliary to the CH system.

In early 1938, Butement started the development of a CD system based on Bowen's evolving 200-MHz (1.5-m) airborne sets. The transmitter had a 400 Hz pulse rate, a 2-μs pulse width, and 50 kW power (later increased to 150 kW). Although many of Bowen's transmitter and receiver components were used, the system would not be airborne so there were no limitations on antenna size.

Primary credit for introducing beamed RDF systems in Britain must be given to Butement. For the CD, he developed a large dipole array, 10 feet (3.0 m) high and 24 feet (7.3 m) wide, giving much narrower beams and higher gain. This could be rotated at a speed up to 1.5 revolutions per minute. For greater directional accuracy, lobe switching on the receiving antennas was adopted. As a part of this development, he formulated the first – at least in Britain – mathematical relationship that would later become well known as the "radar range equation".

By May 1939, the CD RDF could detect aircraft flying as low as 500 feet (150 m) and at a range of 25 mi (40 km). With an antenna 60 feet (18 m) above sea level, it could determine the range of a 2,000-ton ship at 24 mi (39 km) and with an angular accuracy of as little as a quarter of a degree.

Royal Navy

Although the Royal Navy maintained close contact with the Air Ministry work at Bawdsey, they chose to establish their own RDF development at the Experimental Department of His Majesty's Signal School (HMSS) in Portsmouth, Hampshire, on the south coast.

HMSS started RDF work in September 1935. Initial efforts, under R. F. Yeo, were in wavelengths ranging between 75 MHz (4 m) and 1.2 GHz (25 cm). All of the work was under the utmost secrecy; it could not even be discussed with other scientists and engineers at Portsmouth. A 75 MHz range-only set was eventually developed and designated Type 79X. Basic tests were done using a training ship, but the operation was unsatisfactory.

In August 1937, the RDF development at HMSS changed, with many of their best researchers brought into the activity. John D. S. Rawlinson was made responsible for improving the Type 79X. To increase the efficiency, he decreased the frequency to 43 MHz (7 m). Designated Type 79Y, it had separate, stationary transmitting and receiving antennas.

Prototypes of the Type 79Y air-warning system were successfully tested at sea in early 1938. The detection range on aircraft was between 30 and 50 mi (48 and 80 km), depending on height. The systems were then placed into service in August on the cruiser HMS *Sheffield* and in October on the battleship HMS *Rodney*. These were the first vessels in the Royal Navy with RDF systems.

Germany

A radio-based device for remotely indicating the presence of ships was built in Germany by Christian Hülsmeyer in 1904. Often referred to as the first radar system, this did not directly measure the range (distance) to the target, and thus did not meet the criteria to be given this name.

Over the following three decades in Germany, a number of radio-based detection systems were developed but none were true radars. This situation changed before World War II. Developments in three leading industries are described.

GEMA

In the early 1930s, physicist Rudolf Kühnhold, Scientific Director at the *Kriegsmarine* (German navy) *Nachrichtenmittel-Versuchsanstalt* (NVA—Experimental Institute of Communication Systems) in Kiel, was attempting to improve the acoustical methods of underwater detection of ships. He concluded that the desired accuracy in measuring distance to targets could be attained only by using pulsed electromagnetic waves.

During 1933, Kühnhold first attempted to test this concept with a transmitting and receiving set that operated in the microwave region at 13.5 cm (2.22 GHz). The transmitter used a Barkhausen-Kurz tube (the first microwave generator) that produced only 0.1 watt. Unsuccessful with this, he asked for assistance from Paul-Günther Erbslöh and Hans-Karl Freiherr von Willisen, amateur radio operators who were developing a VHF system for communications. They enthusiastically agreed, and in January 1934, formed a company, *Gesellschaft für Elektroakustische und Mechanische Apparate* (GEMA), for the effort. From the start, the firm was always called simply GEMA.

Work on a *Funkmessgerät für Untersuchung* (radio measuring device for research) began in earnest at GEMA. Hans Hollmann and Theodor Schultes, both affiliated with the prestigious Heinrich Hertz Institute in Berlin, were added as consultants. The first apparatus used a split-anode magnetron purchased from Philips in the Netherlands. This provided about 70 W at 50 cm (600 MHz), but suffered from frequency instability. Hollmann built a regenerative receiver and Schultes developed Yagi antennas for transmitting and receiving. In June 1934, large vessels passing through the Kiel Harbor were detected by Doppler-beat interference at a distance of about 2 km (1.2 mi). In October, strong reflections were observed from an aircraft that happened to fly through the beam; this opened consideration of targets other than ships.

Kühnhold then shifted the GEMA work to a pulse-modulated system. A new 50 cm (600 MHz) Philips magnetron with better frequency stability was used. It was modulated with 2- μs pulses at a PRF of 2000 Hz. The transmitting antenna was an array of 10 pairs of dipoles with a reflecting mesh. The wide-band regenerative receiver used Acorn tubes from RCA, and the receiving antenna had three pairs of dipoles and incorporated lobe switching. A blocking device (a duplexer), shut the receiver input when the transmitter pulsed. A Braun tube (a CRT) was used for displaying the range.

The equipment was first tested at a NVA site at the Lübecker Bay near Pelzerhaken. During May 1935, it detected returns from woods across the bay at a range of 15 km (9.3 mi). It had limited success, however, in detecting a research ship, *Welle*, only a short distance away. The receiver was then rebuilt, becoming a super-regenerative set with two intermediate-frequency stages. With this improved receiver, the system readily tracked vessels at up to 8 km (5.0 mi) range.

In September 1935, a demonstration was given to the Commander-in-Chief of the *Kriegsmarine*. The system performance was excellent; the range was read off the Braun tube with a tolerance of 50 meters (less than 1 percent variance), and the lobe switching allowed a directional accuracy of 0.1 degree. Historically, this marked the first naval vessel equipped with radar. Although this apparatus was not put into production, GEMA was funded to develop similar systems operating around 50 cm (500 MHz). These became the Seetakt for the *Kriegsmarine* and the Freya for the *Luftwaffe* (German Air Force).

Kühnhold remained with the NVA, but also consulted with GEMA. He is considered by many in Germany as the Father of Radar. During 1933-6, Hollmann wrote the first comprehensive treatise on microwaves, *Physik und Technik der ultrakurzen Wellen* (Physics and Technique of Ultrashort Waves), Springer 1938.

Telefunken

In 1933, when Kühnhold at the NVA was first experimenting with microwaves, he had sought information from Telefunken on microwave tubes. (Telefunken was the largest supplier of radio products in Germany) There, Wilhelm Tolmé Runge had told him that no vacuum tubes were available for these frequencies. In fact, Runge was already experimenting with high-frequency transmitters and had Telefunken's tube department working on cm-wavelength devices.

In the summer of 1935, Runge, now Director of Telefunken's Radio Research Laboratory, initiated an internally funded project in radio-based detection. Using Barkhausen-Kurz tubes, a 50 cm (600 MHz) receiver and 0.5-W transmitter were built. With the antennas placed flat on the ground some distance apart, Runge arranged for an aircraft to fly overhead and found that the receiver gave a strong Doppler-beat interference signal.

Runge, now with Hans Hollmann as a consultant, continued in developing a 1.8 m (170 MHz) system using pulse-modulation. Wilhelm Stepp developed a transmit-receive device (a duplexer) for allowing a common antenna. Stepp also code-named the system *Darmstadt* after his home town, starting the practice in Telefunken of giving the systems names of cities. The system, with only a few watts transmitter power, was first tested in February 1936, detecting an aircraft at about 5 km (3.1 mi) distance. This led the *Luftwaffe* to fund the development of a 50 cm (600 MHz) gun-laying system, the *Würzburg*.

Lorenz

Since before the First World War, Standard Elektrik Lorenz had been the main supplier of communication equipment for the German military and was the main rival of Telefunken. In late 1935, when Lorenz found that Runge at Telefunken was doing research in radio-based detection equipment, they started a similar activity under Gottfried Müller. A pulse-modulated set called *Einheit für Abfragung* (DFA - Device for Detection) was built. It used a type DS-310 tube (similar to the Acorn) operating at 70 cm (430 MHz) and about 1 kW power, it had identical transmitting and receiving antennas made with rows of half-wavelength dipoles backed by a reflecting screen.

In early 1936, initial experiments gave reflections from large buildings at up to about 7 km (4.3 mi). The power was doubled by using two tubes, and in mid-1936, the equipment was set up on cliffs near Kiel, and good detections of ships at 7 km (4.3 mi) and aircraft at 4 km (2.5 mi) were attained.

The success of this experimental set was reported to the *Kriegsmarine*, but they showed no interest; they were already fully engaged with GEMA for similar equipment. Also, because of extensive agreements between Lorenz and many foreign countries, the naval authorities had reservations concerning the company handling classified work. The DFA was then demonstrated to the *Heer* (German Army), and they contracted with Lorenz for developing *Kurfürst* (Elector), a system for supporting *Flugzeugabwehrkanone* (Flak, anti-aircraft guns).

United States

In the United States, both the Navy and Army needed means of remotely locating enemy ships and aircraft. In 1930, both services initiated the development of radio equipment that could meet this need. There was little coordination of these efforts; thus, they will be described separately.

United States Navy

In the autumn of 1922, Albert H. Taylor and Leo C. Young at the U.S. Naval Aircraft Radio Laboratory were conducting communication experiments when they noticed that a wooden ship in the Potomac River was interfering with their signals. They prepared a memorandum suggesting that this might be used for ship detection in a harbor defense, but their suggestion was not taken up. In 1930, Lawrence A. Hyland working with Taylor and Young, now at the U.S. Naval Research Laboratory (NRL) in Washington, D.C., used a similar arrangement of radio equipment to detect a passing aircraft. This led to a proposal and patent for using this technique for detecting ships and aircraft.

A simple wave-interference apparatus can detect the presence of an object, but it cannot determine its location or velocity. That had to await the invention of pulsed radar, and later, additional encoding techniques to extract this information from a CW signal. When Taylor's group at the NRL were unsuccessful in getting interference radio accepted as a detection means, Young suggested trying pulsing techniques. This would also allow the direct determination of range to the target. In 1924, Hyland and Young had built such a transmitter for Gregory Breit and Merle A. Tuve at the Carnegie Institution of Washington for successfully measuring the height of the ionosphere.

Robert Morris Page was assigned by Taylor to implement Young's suggestion. Page designed a transmitter operating at 60 MHz and pulsed 10 μs in duration and 90 μs between pulses. In December 1934, the apparatus was used to detect a plane at a distance of one mile (1.6 km) flying up and down the Potomac. Although the detection range was small and the indications on the oscilloscope monitor were almost indistinct, it demonstrated the basic concept of a pulsed radar system. Based on this, Page, Taylor, and Young are usually credited with building and demonstrating the world's first true radar.

An important subsequent development by Page was the duplexer, a device that allowed the transmitter and receiver to use the same antenna without overwhelming or destroying the sensitive receiver circuitry. This also solved the problem associated with synchronization of separate transmitter and receiver antennas which is critical to accurate position determination of long range targets.

The experiments with pulsed radar were continued, primarily in improving the receiver for handling the short pulses. In June 1936, the NRL's first prototype radar system, now operating at 28.6 MHz, was demonstrated to government officials, successfully tracking an aircraft at distances up to 25 miles (40 km). Their radar was based on low frequency signals, at least by today's standards, and thus required large antennas, making it impractical for ship or aircraft mounting.

Antenna size is inversely proportional to the operating frequency; therefore, the operating frequency of the system was increased to 200 MHz, allowing much smaller antennas. The frequency of 200 MHz was the highest possible with existing transmitter tubes and other components. The new system was successfully tested at the NRL in April 1937, That same month, the first sea-borne testing was conducted. The equipment was temporarily installed on the USS *Leary*, with a Yagi antenna mounted on a gun barrel for sweeping the field of view.

Based on success of the sea trials, the NRL further improved the system. Page developed the ring oscillator, allowing multiple output tubes and increasing the pulse-power to 15 kW in 5-μs pulses. A 20-by-23 ft (6 x 7 m), stacked-dipole "bedspring" antenna was used. In laboratory test during 1938, the system, now designated XAF, detected planes at ranges up to 100 miles (160 km). It was installed on the battleship USS *New York* for sea trials starting in January 1939, and became the first operational radio detection and ranging set in the U.S. fleet.

In May 1939, a contract was awarded to RCA for production. Designated CXAM, deliveries started in May 1940. The acronym RADAR was coined from "Radio Detection And Ranging". One of the first CXAM systems was placed aboard the USS *California*, a battleship that was sunk in the Japanese attack on Pearl Harbor on December 7, 1941.

United States Army

As the Great Depression started, economic conditions led the U.S. Army Signal Corps to consolidate its widespread laboratory operations to Fort Monmouth, New Jersey. On June 30, 1930, these were designated the Signal Corps Laboratories (SCL) and Lt. Colonel (Dr.) William R. Blair was appointed the SCL Director.

Among other activities, the SCL was made responsible for research in the detection of aircraft by acoustical and infrared radiation means. Blair had performed his doctoral research in the interaction of electromagnet waves with solid materials, and naturally gave attention to this type of detection. Initially, attempts were made to detect infrared radiation, either from the heat of aircraft engines or as reflected from large searchlights with infrared filters, as well as from radio signals generated by the engine ignition.

Some success was made in the infrared detection, but little was accomplished using radio. In 1932, progress at the Naval Research Laboratory (NRL) on radio interference for aircraft detection was passed on to the Army. While it does not appear that any of this information was used by Blair, the SCL did undertake a systematic survey of what was then known throughout the world about the methods of generating, modulating, and detecting radio signals in the microwave region.

The SCL's first definitive efforts in radio-based target detection started in 1934 when the Chief of the Army Signal Corps, after seeing a microwave demonstration by RCA, suggested that radio-echo techniques be investigated. The SCL called this technique radio position-finding (RPF). Based on the previous investigations, the SCL first tried microwaves. During 1934 and 1935, tests of microwave RPF equipment resulted in Doppler-shifted signals being obtained, initially at only a few hundred feet distance and later greater than a mile. These tests involved a bi-static arrangement, with the transmitter at one end of the signal path and the receiver at the other, and the reflecting target passing through or near the path.

Blair was evidently not aware of the success of a pulsed system at the NRL in December 1934. In an internal 1935 note, Blair had commented:

Consideration is now being given to the scheme of projecting an interrupted sequence of trains of oscillations against the target and attempting to detect the echoes during the interstices between the projections.

In 1936, W. Delmar Hershberger, SCL's Chief Engineer at that time, started a modest project in pulsed microwave transmission. Lacking success with microwaves, Hershberger visited the NRL (where he had earlier worked) and saw a demonstration of their pulsed set. Back at the SCL, he and Robert H. Noyes built an experimental apparatus using a 75 watt, 110 MHz (2.73 m) transmitter with pulse modulation and a receiver patterned on the one at the NRL. A request for project funding was turned down by the War Department, but $75,000 for support was diverted from a previous appropriation for a communication project.

In October 1936, Paul E. Watson became the SCL Chief Engineer and led the project. A field setup near the coast was made with the transmitter and receiver separated by a mile. On December 14, 1936, the experimental set detected at up to 7 mi (11 km) range aircraft flying in and out of New York City.

Work then began on a prototype system. Ralph I. Cole headed receiver work and William S. Marks lead transmitter improvements. Separate receivers and antennas were used for azimuth and elevation detection. Both receiving and the transmitting antennas used large arrays of dipole wires on wooden frames. The system output was intended to aim a searchlight. The first demonstration of the full set was made on the night of May 26, 1937. A bomber was detected and then illuminated by the searchlight. The observers included the Secretary of War, Henry A. Woodring; he was so impressed that the next day orders were given for the full development of the system. Congress gave an appropriation of $250,000.

The frequency was increased to 200 MHz (1.5 m). The transmitter used 16 tubes in a ring oscillator circuit (developed at the NRL), producing about 75 kW peak power. Major James C. Moore was assigned to head the complex electrical and mechanical design of lobe switching antennas. Engineers from Western Electric and Westinghouse were brought in to assist in the overall development. Designated SCR-268, a prototype was successfully demonstrated in late 1938 at Fort Monroe, Virginia. The production of SCR-268 sets was started by Western Electric in 1939, and it entered service in early 1941.

Even before the SCR-268 entered service, it had been greatly improved. In a project led by Major (Dr.) Harold A. Zahl, two new configurations evolved – the SCR-270 (mobile) and the SCR-271 (fixed-site). Operation at 106 MHz (2.83 m) was selected, and a single water-cooled tube provided 8 kW (100 kW pulsed) output power. Westinghouse received a production contract, and started deliveries near the end of 1940.

The Army deployed five of the first SCR-270 sets around the island of Oahu in Hawaii. At 7:02 on the morning of December 7, 1941, one of these radars detected a flight of aircraft at a range of 136 miles (219 km) due north. The observation was passed on to an aircraft warning center where it was misidentified as a flight of U.S. bombers known to be approaching from the mainland. The alarm went unheeded, and at 7:48, the Japanese aircraft first struck at Pearl Harbor.

USSR

In 1895, Alexander Stepanovich Popov, a physics instructor at the Imperial Russian Navy school in Kronstadt, developed an apparatus using a coherer tube for detecting distant lightning strikes. The next year, he added a spark-gap transmitter and demonstrated the first radio communication set in Russia. During 1897, while testing this in communicating between two ships in the Baltic Sea, he took note of an interference beat caused by the passage of a third vessel. In his report, Popov wrote that this phenomenon might be used for detecting objects, but he did nothing more with this observation.

In a few years following the 1917 Russian Revolution and the establishment the Union of Soviet Socialist Republics (USSR or Soviet Union) in 1924, Germany's *Luftwaffe* had aircraft capable of penetrating deep into Soviet territory. Thus, the detection of aircraft at night or above clouds was of great interest to the Soviet Air Defense Forces (PVO).

The PVO depended on optical devices for locating targets, and had physicist Pavel K. Oshchepkov conducting research in possible improvement of these devices. In June 1933, Oshchepkov changed his research from optics to radio techniques and started the development of a *razvedyvlatl'naya elektromagnitnaya stantsiya* (reconnaissance electromagnetic station). In a short time, Oshchepkov was made responsible for a technical expertise sector of PVO devoted to *radiolokatory* (radio-location) techniques as well as heading a Special Design Bureau (SKB, spetsialnoe konstruktorskoe byuro) in Leningrad.

Radio-location Beginnings

The *Glavnoe Artilleriyskoe Upravlenie* (GAU, Main Artillery Administration) was considered the "brains" of the Red Army. It not only had competent engineers and physicists on its central staff, but also had a number of scientific research institutes. Thus, the GAU was also assigned the aircraft detection problem, and Lt. Gen. M. M. Lobanov was placed in charge.

After examining existing optical and acoustical equipment, Lobanov also turned to radio-location techniques. For this he approached the *Tsentral'naya Radiolaboratoriya* (TsRL, Central Radio Laboratory) in Leningrad. Here, Yu. K. Korovin was conducting research on VHF communications, and had built a 50 cm (600 MHz), 0.2 W transmitter using a Barkhausen-Kurz tube. For testing the concept, Korovin arranged the transmitting and receiving antennas along the flight path of an aircraft. On January 3, 1934, a Doppler signal was received by reflections from the aircraft at some 600 m range and 100–150 m altitude.

For further research in detection methods, a major conference on this subject was arranged for the PVO by the Russian Academy of Sciences (RAN). The conference was held in Leningrad in mid-January 1934, and chaired by Abram Fedorovich Ioffe, Director of the Leningrad Physical-Technical Institute (LPTI). Ioffe was generally considered the top Russian physicist of his time. All types of detection techniques were discussed, but radio-location received the greatest attention.

To distribute the conference findings to a wider audience, the proceedings were published the following month in a journal. This included all of the then-existing information on radio-location in the USSR, available (in Russian language) to researchers in this field throughout the world.

Recognizing the potential value of radio-location to the military, the GAU made a separate agreement with the Leningrad Electro-Physics Institute (LEPI), for a radio-location system. This technical effort was led by B. K. Shembel. The LEPI had built a transmitter and receiver to study the radio-reflection characteristics of various materials and targets. Shemlbel readily made this into an experimental bi-static radio-location system called *Bistro* (Rapid).

The *Bistro* transmitter, operating at 4.7 m (64 MHz), produced near 200 W and was frequency-modulated by a 1 kHz tone. A fixed transmitting antenna gave a broad coverage of what was called a *radioekran* (radio screen). A regenerative receiver, located some distance from the transmitter, had a dipole antenna mounted on a hand-driven reciprocating mechanism. An aircraft passing into the screened zone would reflect the radiation, and the receiver would detect the Doppler-interference beat between the transmitted and reflected signals.

Bistro was first tested during the summer of 1934. With the receiver up to 11 km away from the transmitter, the set could only detect an aircraft entering a screen at about 3 km (1.9 mi) range and under 1,000 m. With improvements, it was believed to have a potential range of 75 km, and five sets were ordered in October for field trials. *Bistro* is often cited as the USSR's first radar system; however, it was incapable of directly measuring range and thus could not be so classified.

LEPI and TsRL were both made a part of *Nauchno-issledovatelsky institut-9* (NII-9, Scientific Research Institute #9), a new GAU organization opened in Leningrad in 1935. Mikhail A. Bonch-Bruyevich, a renowned radio physicist previously with TsRL and the University of Leningrad, was named the NII-9 Scientific Director.

Research on magnetrons began at Kharkov University in Ukraine during the mid-1920s. Before the end of the decade this had resulted in publications with worldwide distribution, such as the German journal *Annalen der Physik* (*Annals of Physics*). Based on this work, Ioffe recommended that a portion of the LEPI be transferred to the city of Kharkov, resulting in the Ukrainian Institute of Physics and Technology (LIPT) being formed in 1930. Within the LIPT, the Laboratory of Electromagnetic Oscillations (LEMO), headed by Abram A. Slutskin, continued with magnetron development. Led by Aleksandr S. Usikov, a number of advanced segmented-anode magnetrons evolved. (It is noted that these and other early magnetrons developed in the USSR suffered from frequency instability, a problem in their use in Soviet radar systems.)

In 1936, one of Usikov's magnetrons producing about 7 W at 18 cm (1.7 GHz) was used by Shembel at the NII-9 as a transmitter in a *radioiskatel* (radio-seeker) called *Burya* (Storm). Operating similarly to *Bistro*, the range of detection was about 10 km, and provided azimuth and elevation coordinates estimated to within 4 degrees. No attempts were made to make this into a pulsed system, thus, it could not provide range and was not qualified to be classified as a radar. It was, however, the first microwave radio-detection system.

While work by Shembel and Bonch-Bruyevich on continuous-wave systems was taking place at NII-9, Oshehepkov at the SKB and V. V. Tsimbalin of Ioffe's LPTI were pursuing a pulsed system. In 1936, they built a radio-location set operating at 4 m (75 MHz) with a peak-power of about 500 W and a 10-μs pulse duration. Before the end of the year, tests using separated transmitting and receiving sites resulted in an aircraft being detected at 7 km. In April 1937, with the peak-pulse power increased to 1 kW and the antenna separation also increased, test showed a detection range of near 17 km at a height of 1.5 km. Although a pulsed system, it was not capable of directly providing range – the technique of using pulses for determining range had not yet been developed.

Pre-war Radio Location Systems

In June 1937, all of the work in Leningrad on radio-location suddenly stopped. The infamous Great Purge of dictator Joseph Stalin swept over the military high commands and its supporting scientific community. The PVO chief was executed. Oshchepkov, charged with "high crime", was sentenced to 10 years at a Gulag penal labor camp. NII-9 as an organization was saved, but Shenbel was dismissed and Bonch-Bruyevich was named the new director.

The *Nauchnoissledovatel'skii ispytalel'nyi institut svyazi RKKA* (NIIIS-KA, Scientific Research Institute of Signals of the Red Army), had initially opposed research in radio-location, favoring instead acoustical techniques. However, this portion of the Red Army gained power as a result of the Great Purge, and did an about face, pressing hard for speedy development of radio-location systems. They took over Oshchepkov's laboratory and were made responsible for all existing and future agreements for research and factory production. Writing later about the Purge and subsequent effects, General Lobanov commented that it led to the development being placed under a single organization, and the rapid reorganization of the work.

At Oshchepkov's former laboratory, work with the 4 m (75 MHz) pulsed-transmission system was continued by A. I. Shestako. Through pulsing, the transmitter produced a peak power of 1 kW, the highest level thus far generated. In July 1938, a fixed-position, bi-static experimental system detected an aircraft at about 30 km range at heights of 500 m, and at 95 km range, for high-flying targets at 7.5 km altitude. The system was still incapable of directly determining the range. The project was then taken up by Ioffe's LPTI, resulting in the development of a mobile system designated *Redut* (Redoubt). An arrangement of new transmitter tubes was used, giving near 50 kW peak-power with a 10 μs pulse-duration. Yagi antennas were adopted for both transmitting and receiving.

The *Redut* was first field tested in October 1939, at a site near Sevastopol, a port in Ukraine on the coast of the Black Sea. This testing was in part to show the NKKF (Soviet Navy) the value of early-warning radio-location for protecting strategic ports. With the equipment on a cliff about 160 meters above sea level, a flying boat was detected at ranges up to 150 km. The Yagi antennas were spaced about 1,000 meters; thus, close coordination was required to aim them in synchronization. An improved version of the Redut, the Redut-K, was developed by Aksel Berg in 1940 and placed aboard the light cruiser *Molotov* in April 1941. *Molotov* became the first Soviet warship equipped with radar.

At the NII-9 under Bonch-Bruyevich, scientists developed two types of very advanced microwave generators. In 1938, a linear-beam, velocity-modulated vacuum tube (a klystron) was developed by Nikolay Devyatkov, based on designs from Kharkiv. This device produced about 25 W at 15–18 cm (2.0–1.7 GHz) and was later used in experimental systems. Devyatkov followed this with a simpler, single-resonator device (a reflex klystron). At this same time, D. E. Malyarov and N. F. Alekseyev were building a series of magnetrons, also based on designs from Kharkov; the best of these produced 300 W at 9 cm (3 GHz).

Also at NII-9, D. S. Stogov was placed in charge of the improvements to the *Bistro* system. Redesignated as *Reven* (Rhubarb), it was tested in August 1938, but was only marginally better than the predecessor. With additional minor operational improvements, it was made into a mobile system called *Radio Ulavlivatel Samoletov* (RUS, Radio Catcher of Aircraft), soon designated as *RUS-1*. This continuous-wave, bi-static system had a truck-mounted transmitter operating at 4.7 m (64 MHz) and two truck-mounted receivers.

Although the *RUS-1* transmitter was in a cabin on the rear of a truck, the antenna had to be strung between external poles anchored to the ground. A second truck carrying the electrical generator and other equipment was backed against the transmitter truck. Two receivers were used, each in a truck-mounted cabin with a dipole antenna on a rotatable pole extended overhead. In use, the receiver trucks were placed about 40 km apart; thus, with two positions, it would be possible to make a rough estimate of the range by triangulation on a map.

The *RUS-1* system was tested and put into production in 1939, then entered service in 1940, becoming the first deployed radio-location system in the Red Army. About 45 *RUS-1* systems were built at the Svetlana Factory in Leningrad before the end of 1941, and deployed along the western USSR borders and in the Far East. Without direct ranging capability, however, the military found the *RUS-1* to be of little value.

Even before the demise of efforts in Leningrad, the NIIIS-KA had contracted with the UIPT in Kharkov to investigate a pulsed radio-location system for anti-aircraft applications. This led the LEMO, in March 1937, to start an internally funded project with the code name *Zenit* (a popular football team at the time). The transmitter development was led by Usikov, supplier of the magnetron used earlier in the *Burya*. For the *Zenit*, Usikov used a 60 cm (500 MHz) magnetron pulsed at 10–20 μs duration and providing 3 kW pulsed power, later increased to near 10 kW. Semion Braude led the development of a superheterodyne receiver using a tunable magnetron as the local oscillator. The system had separate transmitting and receiving antennas set about 65 m apart, built with dipoles backed by 3-meter parabolic reflectors.

Zenit was first tested in October 1938. In this, a medium-sized bomber was detected at a range of 3 km. The testing was observed by the NIIIS-KA and found to be sufficient for starting a contracted effort. An agreement was made in May 1939, specifying the required performance and calling for the system to be ready for production by 1941. The transmitter was increased in power, the antennas had selsens added to allow them to track, and the receiver sensitivity was improved by using an RCA 955 acorn triode as the local oscillator.

A demonstration of the improved *Zenit* was given in September 1940. In this, it was shown that the range, altitude, and azimuth of an aircraft flying at heights between 4,000 and 7,000 meters could be determined at up to 25 km distance. The time required for these measurements, however, was about 38 seconds, far too long for use by anti-aircraft batteries. Also, with the antennas aimed at a low angle, there was a dead zone of some distance caused by interference from ground-level reflections. While this performance was not satisfactory for immediate gun-laying applications, it was the first full three-coordinate radio-location system in the Soviet Union and showed the way for future systems.

Work at the LEMO continued on *Zenit*, particularly in converting it into a single-antenna system designated *Rubin*. This effort, however, was disrupted by the invasion of the USSR by Germany in June 1941. In a short while, the development activities at Kharkov were ordered to be evacuated to the Far East. The research efforts in Leningrad were similarly dispersed.

After eight years of effort by highly qualified physicists and engineers, the USSR entered World War II without a fully developed and fielded radar system.

Japan

As a seafaring nation, Japan had an early interest in wireless (radio) communications. The first known use of wireless telegraphy in warfare at sea was by the Imperial Japanese Navy, in defeating the Russian Imperial Fleet in 1904 at the Battle of Port Arthur. There was an early interest in equipment for radio direction-finding, for use in both navigation and military surveillance. The Imperial Navy developed an excellent receiver for this purpose in 1921, and soon most of the Japanese warships had this equipment.

In the two decades between the two World Wars, radio technology in Japan made advancements on a par with that in the western nations. There were often impediments, however, in transferring these advancements into the military. For a long time, the Japanese had believed that they had the best fighting capability of any military force in the world. The military leaders, who were then also in control of the government, sincerely felt that the weapons, aircraft, and ships that they had built were fully sufficient and, with these as they were, the Japanese Army and Navy were invincible. In 1936, Japan joined Nazi Germany and Fascist Italy in a Tripartite Pact.

Technology Background

Radio engineering was strong in Japan's higher education institutions, especially the Imperial (government-financed) universities. This included undergraduate and graduate study, as well as academic research in this field. Special relationships were established with foreign universities and institutes, particularly in Germany, with Japanese teachers and researchers often going overseas for advanced study.

The academic research tended toward the improvement of basic technologies, rather than their specific applications. There was considerable research in high-frequency and high-power oscillators, such as the magnetron, but the application of these devices was generally left to industrial and military researchers.

One of Japan's best-known radio researchers in the 1920s-1930s era was Professor Hidetsugu Yagi. After graduate study in Germany, England, and America, Yagi joined Tohoku University, where his research centered on antennas and oscillators for high-frequency communications. A summary of the radio research work at Tohoku University was contained in a 1928 seminal paper by Yagi.

Jointly with Shintaro Uda, one of Yagi's first doctoral students, a radically new antenna emerged. It had a number of parasitic elements (directors and reflectors) and would come to be known as the Yagi-Uda or Yagi antenna. A U.S. patent, issued in May 1932, was assigned to RCA. To this day, this is the most widely used directional antenna worldwide.

The cavity magnetron was also of interest to Yagi. This HF (~10-MHz) device had been invented in 1921 by Albert W. Hull at General Electric, and Yagi was convinced that it could function in the VHF or even the UHF region. In 1927, Kinjiro Okabe, another of Yagi's early doctoral students, developed a split-anode device that ultimately generated oscillations at wavelengths down to about 12 cm (2.5 GHz).

Researchers at other Japanese universities and institutions also started projects in magnetron development, leading to improvements in the split-anode device. These included Kiyoshi Morita at the Tokyo Institute of Technology, and Tsuneo Ito at Tokoku University.

Shigeru Nakajima at Japan Radio Company (JRC) saw a commercial potential of these devices and began the further development and subsequent very profitable production of magnetrons for the medical dielectric heating (diathermy) market. The only military interest in magnetrons was shown by Yoji Ito at the Naval Technical Research Institute (NTRI).

The NTRI was formed in 1922, and became fully operational in 1930. Located at Meguro, Tokyo, near the Tokyo Institute of Technology, first-rate scientists, engineers, and technicians were engaged in activities ranging from designing giant submarines to building new radio tubes. Included were all of the precursors of radar, but this did not mean that the heads of the Imperial Navy accepted these accomplishments.

In 1936, Tsuneo Ito (no relationship to Yoji Ito) developed an 8-split-anode magnetron that produced about 10 W at 10 cm (3 GHz). Based on its appearance, it was named *Tachibana* (or Mandarin, an orange citrus fruit). Tsuneo Ito also joined the NTRI and continued his research on magnetrons in association with Yoji Ito. In 1937, they developed the technique of coupling adjacent segments (called push-pull), resulting in frequency stability, an extremely important magnetron breakthrough.

By early 1939, NTRI/JRC had jointly developed a 10-cm (3-GHz), stable-frequency Mandarin-type magnetron (No. M3) that, with water cooling, could produce 500-W power. In the same time period, magnetrons were built with 10 and 12 cavities operating as low as 0.7 cm (40 GHz). The configuration of the M3 magnetron was essentially the same as that used later in the magnetron developed by Boot and Randall at Birmingham University in early 1940, including the improvement of strapped cavities. Unlike the high-power magnetron in Britain, however, the initial device from the NTRI generated only a few hundred watts.

In general, there was no lack of scientific and engineering capabilities in Japan; their warships and aircraft clearly showed high levels of technical competency. They were ahead of Britain in the development of magnetrons, and their Yagi antenna was the world standard for VHF systems. It was simply that the top military leaders failed to recognize how the application of radio in detection and ranging – what was often called the Radio Range Finder (RRF) – could be of value, particularly in any defensive role; offense not defense, totally dominated their thinking.

Imperial Army

In 1938, engineers from the Research Office of Nippon Electric Company (NEC) were making coverage tests on high-frequency transmitters when rapid fading of the signal was observed. This occurred whenever an aircraft passed over the line between the transmitter and receiving meter. Masatsugu Kobayashi, the Manager of NEC's Tube Department, recognized that this was due to the beat-frequency interference of the direct signal and the Doppler-shifted signal reflected from the aircraft.

Kobayashi suggested to the Army Science Research Institute that this phenomenon might be used as an aircraft warning method. Although the Army had rejected earlier proposals for using radio-detection techniques, this one had appeal because it was based on an easily understandable method and would require little developmental cost and risk to prove its military value. NEC assigned Kinji Satake of their Research Institute to develop a system called the Bi-static Doppler Interference Detector (BDID).

For testing the prototype system, it was set up on an area recently occupied by Japan along the coast of China. The system operated between 4.0-7.5 MHz (75–40 m) and involved a number of widely spaced stations; this formed a radio screen that could detect the presence (but nothing more) of an aircraft at distances up to 500 km (310 mi). The BDID was the Imperial Army's first deployed radio-based detection system, placed into operation in early 1941.

A similar system was developed by Satake for the Japanese homeland. Information centers received oral warnings from the operators at BDID stations, usually spaced between 65 and 240 km (40 and 150 mi). To reduce homing vulnerability – a great fear of the military – the transmitters operated with only a few watts power. Although originally intended to be temporary until better systems were available, they remained in operation throughout the war. It was not until after the start of war that the Imperial Army had equipment that could be called radar.

Imperial Navy

In the mid-1930s, some of the technical specialists in the Imperial Navy became interested in the possibility of using radio to detect aircraft. For consultation, they turned to Professor Yagi who was the Director of the Radio Research Laboratory at Osaka Imperial University. Yagi suggested that this might be done by examining the Doppler frequency-shift in a reflected signal.

Funding was provided to the Osaka Laboratory for experimental investigation of this technique. Kinjiro Okabe, the inventor of the split-anode magnetron and who had followed Yagi to Osaka, led the effort. Theoretical analyses indicated that the reflections would be greater if the wavelength was approximately the same as the size of aircraft structures. Thus, a VHF transmitter and receiver with Yagi antennas separated some distance were used for the experiment.

In 1936, Okabe successfully detected a passing aircraft by the Doppler-interference method; this was the first recorded demonstration in Japan of aircraft detection by radio. With this success, Okabe's research interest switched from magnetrons to VHF equipment for target detection. This, however, did not lead to any significant funding. The top levels of the Imperial Navy believed that any advantage of using radio for this purpose were greatly outweighed by enemy intercept and disclosure of the sender's presence.

Historically, warships in formation used lights and horns to avoid collision at night or when in fog. Newer techniques of VHF radio communications and direction-finding might also be used, but all of these methods were highly vulnerable to enemy interception. At the NTRI, Yoji Ito proposed that the UHF signal from a magnetron might be used to generate a very narrow beam that would have a greatly reduced chance of enemy detection.

Development of microwave system for collision avoidance started in 1939, when funding was provided by the Imperial Navy to JRC for preliminary experiments. In a cooperative effort involving Yoji Ito of the NTRI and Shigeru Nakajima of JRC, an apparatus using a 3-cm (10-GHz) magnetron with frequency modulation was designed and built. The equipment was used in an attempt to detect reflections from tall structures a few kilometers away. This experiment gave poor results, attributed to the very low power from the magnetron.

The initial magnetron was replaced by one operating at 16 cm (1.9 GHz) and with considerably higher power. The results were then much better, and in October 1940, the equipment obtained clear echoes from a ship in Tokyo Bay at a distance of about 10 km (6.2 mi). There was still no commitment by top Japanese naval officials for using this technology aboard warships. Nothing more was done at this time, but late in 1941, the system was adopted for limited use.

In late 1940, Japan arranged for two technical missions to visit Germany and exchange information about their developments in military technology. Commander Yoji Ito represented the Navy's interest in radio applications, and Lieutenant Colonel Kinji Satake did the same for the Army. During a visit of several months, they exchanged significant general information, as well as limited secret materials in some technologies, but little directly concerning radio-detection techniques. Neither side even mentioned magnetrons, but the Germans did apparently disclose their use of pulsed techniques.

After receiving the reports from the technical exchange in Germany, as well as intelligence reports concerning the success of Britain with firing using RDF, the Naval General Staff reversed itself and tentatively accepted pulse-transmission technology. On August 2, 1941, even before Yoji Ito returned to Japan, funds were allocated for the initial development of pulse-modulated radars. Commander Chuji Hashimoto of the NTRI was responsible for initiating this activity.

A prototype set operating at 4.2 m (71 MHz) and producing about 5 kW was completed on a crash basis. With the NTRI in the lead, the firm NEC and the Research Laboratory of Japan Broadcasting Corporation (NHK) made major contributions to the effort. Kenjiro Takayanagi, Chief Engineer of NHK's experimental television station and called "the father of Japanese television", was especially helpful in rapidly developing the pulse-forming and timing circuits, as well as the receiver display. In early September 1941, the prototype set was first tested; it detected a single bomber at 97 km (60 mi) and a flight of aircraft at 145 km (90 mi).

The system, Japan's first full Radio Range Finder (RRF – radar), was designated Mark 1 Model 1. Contracts were given to three firms for serial production; NEC built the transmitters and pulse modulators, Japan Victor the receivers and associated displays, and Fuji Electrical the antennas and their servo drives. The system operated at 3.0 m (100 MHz) with a peak-power of 40 kW. Dipole arrays with matte+-type reflectors were used in separate antennas for transmitting and receiving.

In November 1941, the first manufactured RRF was placed into service as a land-based early-warning system at Katsuura, Chiba, a town on the Pacific coast about 100 km (62 mi) from Tokyo. A large system, it weighed close to 8,700 kg (19,000 lb). The detection range was about 130 km (81 mi) for single aircraft and 250 km (160 mi) for groups.

Netherlands

Early radio-based detection in the Netherlands was along two independent lines: one a microwave system at the firm Philips and the other a VHF system at a laboratory of the Armed Forces.

The Philips Company in Eindhoven, Netherlands, operated *Natuurkundig Laboratorium* (NatLab) for fundamental research related to its products. NatLab researcher Klaas Posthumus developed a magnetron split into four elements. In developing a communication system using this magnetron, C.H.J.A. Staal was testing the transmission by using parabolic transmitting and receiving antennas set side-by-side, both aimed at a large plate some distance away. To overcome frequency instability of the magnetron, pulse modulation was used. It was found that the plate reflected a strong signal.

Recognizing the potential importance of this as a detection device, NatLab arranged a demonstration for the *Koninklijke Marine* (Royal Netherlands Navy). This was conducted in 1937 across the entrance to the main naval port at Marsdiep. Reflections from sea waves obscured the return from the target ship, but the Navy was sufficiently impressed to initiate sponsorship of the research. In 1939, an improved set was demonstrated at Wijk aan Zee, detecting a vessel at a distance of 3.2 km (2.0 mi).

A prototype system was built by Philips, and plans were started by the firm Nederlandse Seintoestellen Fabriek (a Philips subsidiary) for building a chain of warning stations to protect the primary ports. Some field testing of the prototype was conducted, but the project was discontinued when Germany invaded the Netherlands on May 10, 1940. Within the NatLab, however, the work was continued in great secrecy until 1942.

During the early 1930s, there were widespread rumours of a "death ray" being developed. The Dutch Parliament set up a Committee for the Applications of Physics in Weaponry under G.J. Elias to examine this potential, but the Committee quickly discounted death rays. The Committee did, however, establish the *Laboratorium voor Fysieke Ontwikkeling* (LFO, Laboratory for Physical Development), dedicated to supporting the Netherlands Armed Forces.

Operating in great secrecy, the LFO opened a facility called the *Meetgebouw* (Measurements Building) located on the Plain of Waalsdorp. In 1934, J.L.W.C. von Weiler joined the LFO and, with S.G. Gratama, began research on a 1.25-m (240-MHz) communication system to be used in artillery spotting.

In 1937, while tests were being conducted on this system, a passing flock of birds disturbed the signal. Realizing that this might be a potential method for detecting aircraft, the Minister of War ordered continuation of the experiments. Weiler and Gratama set about developing a system for directing searchlights and aiming anti-aircraft guns.

The experimental "electrical listening device" operated at 70 cm (430 MHz) and used pulsed transmission at an RPF of 10 kHz. A transmit-receive blocking circuit was developed to allow a common antenna. The received signal was displayed on a CR tube with a circular time base. This set was demonstrated to the Army in April 1938 and detected an aircraft at a range of 18 km (11 mi). The set was rejected, however, because it could not withstand the harsh environment of Army combat conditions.

The Navy was more receptive. Funding was provided for final development, and Max Staal was added to the team. To maintain secrecy, they divided the development into parts. The transmitter was built at the Delft Technical College and the receiver at the University of Leiden. Ten sets would be assembled under the personal supervision of J.J.A. Schagen van Leeuwen, head of the firm Hazemeijer Fabriek van Signaalapparaten.

The prototype had a peak-power of 1 kW, and used a pulse length of 2 to 3 μs with a 10- to 20 kHz PRF. The receiver was a super-heterodyne type using Acorn tubes and a 6 MHz IF stage. The antenna consisted of 4 rows of 16 half-wave dipoles backed by a 3- by 3-meter mesh screen. The operator used a bicycle-type drive to rotate the antenna, and the elevation could be changed using a hand crank.

Several sets were completed, and one was put into operation on the Malieveld in The Hague just before the Netherlands fell to Germany in May 1940. The set worked well, spotting enemy aircraft during the first days of fighting. To prevent capture, operating units and plans for the system were destroyed. Von Weiler and Max Staal fled to England aboard one of the last ships able to leave, carrying two disassembled sets with them. Later, Gratama and van Leeuwen also escaped to England.

France

In 1927, French physicists Camille Gutton and Emile Pierret experimented with magnetrons and other devices generating wavelengths going down to 16 cm. Camille's son, Henri Gutton, was with the *Compagnie Générale de Télégraphie Sans Fil* (CSF) where he and Robert Warneck improved his father's magnetrons.

In 1934, following systematic studies on the magnetron, the research branch of the CSF, headed by Maurice Ponte, submitted a patent application for a device designed to detect obstacles using continuous radiation of ultra-short wavelengths produced by a magnetron. These were still CW systems and depended on Doppler interference for detection. However, as most modern radars, antennas were collocated. The device was measuring distance and azimuth but not directly as in the later "radar" on a screen (1939). Still, this was the first patent of an operational radio-detection apparatus using centimetric wavelengths.

The system was tested in late 1934 aboard the cargo ship *Oregon*, with two transmitters working at 80 cm and 16 cm wavelengths. Coastlines and boats were detected from a range of 10-12 nautical miles. The shortest wavelength was chosen for the final design, which equipped the liner *SS Normandie* as early as mid-1935 for operational use.

In late 1937, Maurice Elie at SFR developed a means of pulse-modulating transmitter tubes. This led to a new 16-cm system with a peak power near 500 W and a pulse width of 6 μs. French and U.S. patents were filed in December 1939. The system was planned to be sea-tested aboard the *Normandie*, but this was cancelled at the outbreak of war.

At the same time, Pierre David at the *Laboratoire National de Radioélectricité* (National Laboratory of Radioelectricity, LNR) experimented with reflected radio signals at about a meter wavelength. Starting in 1931, he observed that aircraft caused interference to the signals. The LNR then initiated research on a detection technique called *barrage électromagnétique* (electromagnetic curtain). While this could indicate the general location of penetration, precise determination of direction and speed was not possible.

In 1936, the *Défense Aérienne du Territoire* (Defence of Air Territory), ran tests on David's electromagnetic curtain. In the tests, the system detected most of the entering aircraft, but too many were missed. As the war grew closer, the need for an aircraft detection was critical. David realized the advantages of a pulsed system, and in October 1938 he designed a 50 MHz, pulse-modulated system with a peak-pulse power of 12 kW. This was built by the firm SADIR.

France declared war on Germany on September 1, 1939, and there was a great need for an early-warning detection system. The SADIR system was taken to near Toulon, and detected and measured the range of invading aircraft as far as 55 km (34 mi). The SFR pulsed system was set up near Paris where it detected aircraft at ranges up to 130 km (81 mi). However, the German advance was overwhelming and emergency measures had to be taken; it was too late for France to develop radars alone and it was decided that her breakthroughs would be shared with her allies.

In mid-1940, Maurice Ponte, from the laboratories of CSF in Paris, presented a cavity magnetron designed by Henri Gutton at SFR to the GEC laboratories at Wembley, Britain. This magnetron was designed for pulsed operation at a wavelength of 16 cm. Unlike other magnetron designs to that day, such as the Boots and Randall magnetron, this tube used an oxide-coated cathode with a peak power output of 1 kW, demonstrating that oxide cathodes were the solution for producing high-power pulses at short wavelengths, a problem which had eluded British and American researchers for years. The significance this event was underlined by Eric Megaw, in a 1946 review of early radar developments: "This was the starting point of the use of the oxide cathode in practically all our subsequent pulsed transmitting waves and as such was a significant contribution to British radar. The date was the 8th May 1940". A tweaked version of this magnetron reached a peak output of 10 kW by August 1940. It was that model which, in turn, was handed to the Americans as a token of good faith during the negotiations made by the Tizard delegation in 1940 to obtain from the U.S. the resources necessary for Britain to exploit the full military potential of her research and development work.

Italy

Guglielmo Marconi initiated the research in Italy on radio-based detection technology. In 1933, while participating with his Italian firm in experiments with a 600 MHz communications link across Rome, he noted transmission disturbances caused by moving objects adjacent to its path. This led to the development at his laboratory at Cornegliano of a 330-MHz (0.91-m) CW Doppler detection system that he called *radioecometro*. Barkhausen-Kurz tubes were used in both the transmitter and receiver.

In May 1935, Marconi demonstrated his system to the Fascist dictator Benito Mussolini and members of the military General Staff; however the output power was insufficient for military use. While Marconi's demonstration raised considerable interest, little more was done with his apparatus.

Mussolini directed that radio-based detection technology be further developed, and it was assigned to the *Regio Instituto Electrotecnico e delle Comunicazioni* (RIEC, Royal Institute for Electro-technics and Communications). The RIEC had been established in 1916 on the campus of the Italian Naval Academy in Livorno. Lieutenant Ugo Tiberio, a physics and radio-technology instructor at the Academy, was assigned to head the project on a part-time basis.

Tiberio prepared a report on developing an experimental apparatus that he called *telemetro radiofonico del rivelatore* (RDT, Radio-Detector Telemetry). The report, submitted in mid-1936, included what was later known as the radar range equation. When the work got underway, Nello Carrara, a civilian physics instructor who had been doing research at the RIEC in microwaves, was added to be responsible for developing the RDT transmitter.

Before the end of 1936, Tiberio and Carrara had demonstrated the EC-1, the first Italian RDT system. This had an FM transmitter operating at 200 MHz (1.5 m) with a single parabolic cylinder antenna. It detected by mixing the transmitted and the Doppler-shifted reflected signals, resulting in an audible tone.

The EC-1 did not provide a range measurement; to add this capability, development of a pulsed system was initiated in 1937. Captain Alfeo Brandimarte joined the group and primarily designed the first pulsed system, the EC-2. This operated at 175 MHz (1.7 m) and used a single antenna made with a number of equi-phased dipoles. The detected signal was intended to be displayed on an oscilloscope. There were many problems, and the system never reached the testing stage.

Work then turned to developing higher power and operating frequencies. Carrara, in cooperation with the firm FIVRE, developed a magnetron-like device. This was composed of a pair of triodes connected to a resonate cavity and produced 10 kW at 425 MHz (70 cm). It was used in designing two versions of the EC-3, one for shipboard and the other for coastal defense.

Italy, joining Germany, entered WWII in June 1940 without an operational RDT. A breadboard of the EC-3 was built and tested from atop a building at the Academy, but most RDT work was stopped as direct support of the war took priority.

Others

In early 1939, the British Government invited representatives from the most technically advanced Commonwealth Nations to visit England for briefings and demonstrations on the highly secret RDF (radar) technology. Based on this, RDF developments were started in Australia, Canada, New Zealand, and South Africa by September 1939. In addition, this technology was independently developed in Hungary early in the war period.

In Australia, the Radiophysics Laboratory was established at Sydney University under the Council for Scientific and Industrial Research; John H. Piddington was responsible for RDF development. The first project was a 200-MHz (1.5-m) shore-defense system for the Australian Army. Designated ShD, this was first tested in September 1941, and eventually installed at 17 ports. Following the Japanese attack on Pearl Harbor, the Royal Australian Air Force urgently needed an air-warning system, and Piddington's team, using the ShD as a basis, put the AW Mark I together in five days. It was being installed in Darwin, Northern Territory, when Australia received the first Japanese attack on February 19, 1942. A short time later, it was converted to a light-weight transportable version, the LW-AW Mark II; this was used by the Australian forces, as well as the U.S. Army, in early island landings in the South Pacific.

The early RDF developments in Canada were at the Radio Section of the National Research Council of Canada. Using commercial components and with essentially no further assistance from Britain, John Tasker Henderson led a team in developing the Night Watchman, a surface-warning system for the Royal Canadian Navy to protect the entrance to the Halifax Harbour. Successfully tested in July 1940, this set operated at 200 MHz (1.5 m), had a 1 kW output with a pulse length of 0.5 μs, and used a relatively small, fixed antenna. This was followed by a ship-borne set designated Surface Warning 1st Canadian (SW1C) with the antenna hand-rotated through the use of a Chevrolet steering wheel in the operator's compartment. The SW1C was first tested at sea in mid-May 1941, but the performance was so poor compared to the Royal Navy's Model 271 ship-borne radar that the Royal Canadian Navy eventually adopted the British 271 in place of the SW1C.

For coastal defense by the Canadian Army, a 200 MHz set with a transmitter similar to the Night Watchman was developed. Designated CD, it used a large, rotating antenna atop a 70-foot (21 m) wooden tower. The CD was put into operation in January 1942.

Ernest Marsden represented New Zealand at the briefings in England, and then established two facilities for RDF development – one in Wellington at the Radio Section of the Central NZ Post Office, and another at Canterbury University College in Christchurch. Charles N. Watson-Munro led the development of land-based and airborne sets at Wellington, while Frederick W. G. White led the development of shipboard sets at Christchurch.

Before the end of 1939, the Wellington group had converted an existing 180-MHz (1.6-m), 1 kW transmitter to produce 2-μs pulses and tested it to detect large vessels at up to 30 km; this was designated CW (Coastal Watching). A similar set, designated CD (Coast Defense) used a CRT for display and had lobe-switching on the receiving antenna; this was deployed in Wellington in late 1940. A partially completed ASV 200 MHz set was brought from Britain by Marsden, and another group at Wellington built this into an aircraft set for the Royal New Zealand Air Force; this was first flown in early 1940. At Christchurch, there was a smaller staff and work went slower, but by July 1940, a 430-MHz (70-cm), 5 kW set was tested. Two types, designated SW (Ship Warning) and SWG (Ship Warning, Gunnery), were placed into service by the Royal New Zealand Navy starting in August 1941. In all some 44 types were developed in New Zealand during WW1.

South Africa did not have a representative at the 1939 meetings in England, but in mid-September, as Ernest Marsden was returning by ship to New Zealand, Basil F. J. Schonland came aboard and received three days of briefings. Schonland, a world authority on lightning and Director of the Bernard Price Institute of Geophysics at Witwatersrand University, immediately started an RDF development using amateur radio components and Institute's lightning-monitoring equipment. Designated JB (for Johannesburg), the 90-MHz (3.3-m), 500-W mobile system was tested in November 1939, just two months after its start. The prototype was operated in Durban before the end of 1939, detecting ships and aircraft at distances up to 80 km, and by the next March a system was fielded by anti-aircraft brigades of the South African Defence Force.

In Hungary, Zoltán Lajos Bay was a Professor of Physics at the Technical University of Budapest as well as the Research Director of Egyesült Izzolampa (IZZO), a radio and electrical manufacturing firm. In late 1942, IZZO was directed by the Minister of Defense to develop a radio-location (*rádiólokáció*, radar) system. Using journal papers on ionospheric measurements for information on pulsed transmission, Bay developed a system called *Sas* (Eagle) around existing communications hardware.

The *Sas* operated at 120 MHz (2.5 m) and was in a cabin with separate transmitting and receiving dipole arrays attached; the assembly was all on a rotatable platform. According to published records, the system was tested in 1944 atop Mount János and had a range of "better than 500 km". A second *Sas* was installed at another location. There is no indication that either *Sas* installation was ever in regular service. After the war, Bay used a modified *Sas* to successfully bounce a signal off the moon.

World War II Radar

At the start of World War II in September 1939, both the United Kingdom and Germany knew of each other's ongoing efforts in radio navigation and its countermeasures – the "Battle of the beams". Also, both nations were generally aware of, and intensely interested in, the other's developments in radio-based detection and tracking, and engaged in an active campaign of espionage and false leaks about their respective equipment. By the time of the Battle of Britain, both sides were deploying range and direction-finding units (radars) and control stations as part of integrated air defense capability. However, the German *Funkmessgerät* (radio measuring device) systems could not assist in an offensive role and was thus not supported by Adolf Hitler. Also, the *Luftwaffe* did not sufficiently appreciate the importance of British Range and Direction Finding (RDF) stations as part of RAF's air defense capability, contributing to their failure.

While the United Kingdom and Germany led in pre-war advances in the use of radio for detection and tracking of aircraft, there were also developments in the United States, the Soviet Union, and Japan. Wartime systems in all of these nations will be summarized. The acronym RADAR (for RAdio Detection And Ranging) was coined by the U.S. Navy in 1940, and the subsequent name "radar" was soon widely used.

When France had just fallen to the Nazis and Britain had no money to develop the magnetron on a massive scale, Churchill agreed that Sir Henry Tizard should offer the magnetron to the Americans in exchange for their financial and industrial help (the Tizard Mission). An early 6 kW version, built in England by the General Electric Company Research Laboratories, Wembley, London, was given to the US government in September 1940. The British magnetron was a thousand times more powerful than the best American transmitter at the time and produced accurate pulses. At the time the most powerful equivalent microwave producer available in the US (a klystron) had a power of only ten watts. The cavity magnetron was widely used during World War II in microwave radar equipment and is often credited with giving Allied radar a considerable performance advantage over German and Japanese radars, thus directly influencing the outcome of the war. It was later described by noted Historian James Phinney Baxter III as "The most valuable cargo ever brought to our shores".

The Bell Telephone Laboratories made a producible version from the magnetron delivered to America by the Tizard Mission, and before the end of 1940, the Radiation Laboratory had been set up on the campus of the Massachusetts Institute of Technology to develop various types of radar using the magnetron. By early 1941, portable centimetric airborne radars were being tested in American and British aircraft. In late 1941, the Telecommunications Research Establishment in Great Britain used the magnetron to develop a revolutionary airborne, ground-mapping radar codenamed H2S. The H2S radar was in part developed by Alan Blumlein and Bernard Lovell. The magnetron radars used by the US and Britain could spot the periscope of a U boat.

Post-war Radar

World War II, which gave impetus to the great surge in radar development, ended between the Allies and Germany in May 1945, followed by Japan in August. With this, radar activities in Germany and Japan ceased for a number of years. In other countries, particularly the United States, Britain, and the USSR, the politically unstable post-war years saw continued radar improvements for military applications. In fact, these three nations all made significant efforts in bringing scientists and engineers from Germany to work in their weapon programs; in the U.S., this was under Operation Paperclip.

Even before the end of the war, various project directed toward non-military applications of radar and closely related technologies were initiated. The US Army Air Forces and the British RAF had made wartime advances in using radar for handling aircraft landing, and this was rapidly expanded into the civil sector. The field of radio astronomy was one of the related technologies; although discovered before the war, it immediately flourished in the late 1940s with many scientists around the world establishing new careers based on their radar experience.

Four techniques, highly important in post-war radars, were matured in the late 1940s-early 1950s: pulse Doppler, monopulse, phased array, and synthetic aperture; the first three were known and even used during wartime developments, but were matured later.

- Pulse-Doppler radar (often known as moving target indication or MTI), uses the Doppler-shifted signals from targets to better detect moving targets in the presence of clutter.
- Monopulse radar (also called simultaneous lobing) was conceived by Robert Page at the NRL in 1943. With this, the system derives error-angle information from a single pulse, greatly improving the tracking accuracy.
- Phased-array radar has the many segments of a large antenna separately controlled, allowing the beam to be quickly directed. This greatly reduces the time necessary to change the beam direction from one point to another, allowing almost simultaneous tracking of multiple targets while maintaining overall surveillance.
- Synthetic-aperture radar (SAR), was invented in the early 1950s at Goodyear Aircraft Corporation. Using a single, relatively small antenna carried on an aircraft, a SAR combines the returns from each pulse to produce a high-resolution image of the terrain comparable to that obtained by a much larger antenna. SAR has wide applications, particularly in mapping and remote sensing.

One of the early applications of digital computers was in switching the signal phase in elements of large phased-array antennas. As smaller computers came into being, these were quickly applied to digital signal processing using algorithms for improving radar performance.

Other advances in radar systems and applications in the decades following WWII are far too many to be included herein. The following sections are intended to provide representative samples.

Military Radars

In the United States, the Rad Lab at MIT officially closed at the end of 1945. The Naval Research Laboratory (NRL) and the Army's Evans Signal Laboratory continued with new activities in centimeter radar development. The United States Air Force (USAF) – separated from the Army in 1946 – concentrated radar research at their Cambridge Research Center (CRC) at Hanscom Field, Massachusetts. In 1951, MIT opened the Lincoln Laboratory for joint developments with the CRC. While the Bell Telephone Laboratories embarked on major communications upgrades, they continued with the Army in radar for their ongoing Nike air-defense program.

In Britain, the RAF's Telecommunications Research Establishment (TRE) and the Army's Radar Research and Development Establishment (RRDE) both continued at reduced levels at Malvern, Worcestershire, then in 1953 were combined to form the Radar Research Establishment. In 1948, all of the Royal Navy's radio and radar R&D activities were combined to form the Admiralty Signal and Radar Establishment, located near Portsmouth, Hampshire. The USSR, although devastated by the war, immediately embarked on the development of new weapons, including radars.

During the Cold War period following WWII, the primary "axis" of combat shifted to lie between the United States and the Soviet Union. By 1949, both sides had nuclear weapons carried by bombers. To provide early warning of an attack, both deployed huge radar networks of increasing sophistication at ever-more remote locations. In the West, the first such system was the Pinetree Line, deployed across Canada in the early 1950s, backed up with radar pickets on ships and oil platforms off the east and west coasts.

The Pinetree Line initially used vintage pulsed radars and was soon supplemented with the Mid-Canada Line (MCL). Soviet technology improvements made these Lines inadequate and, in a construction project involving 25,000 persons, the Distant Early Warning Line (DEW Line) was completed in 1957. Stretching from Alaska to Baffin Island and covering over 6,000 miles (9,700 km), the DEW Line consisted of 63 stations with AN/FPS-19 high-power, pulsed, L-Band radars, most augmented by AN/FPS-23 pulse-Doppler systems. The Soviet Unit tested its first Intercontinental Ballistic Missile (ICBM) in August 1957, and in a few years the early-warning role was passed almost entirely to the more capable DEW Line.

Both the U.S. and the Soviet Union then had ICBMs with nuclear warheads, and each began the development of a major anti-ballistic missile (ABM) system. In the USSR, this was the Fakel V-1000, and for this they developed powerful radar systems. This was eventually deployed around Moscow as the A-35 anti-ballistic missile system, supported by radars designated by NATO as the Cat House, Dog House, and Hen House.

In 1957, the U.S. Army initiated an ABM system first called Nike-X; this passed through several names, eventually becoming the Safeguard Program. For this, there was a long-range Perimeter Acquisition Radar (PAR) and a shorter-range, more precise Missile Site Radar (MSR).

The PAR was housed in a 128-foot (39 m)-high nuclear-hardened building with one face sloping 25 degrees facing north. This contained 6,888 antenna elements separated in transmitting and receiving phased arrays. The L-Band transmitter used 128 long-life traveling-wave tubes (TWTs), having a combined power in the megawatt range The PAR could detect incoming missiles outside the atmosphere at distances up to 1,800 miles (2,900 km).

The MSR had an 80-foot (24 m), truncated pyramid structure, with each face holding a phased-array antenna 13 feet (4.0 m) in diameter and containing 5,001 array elements used for both transmitting and receiving. Operating in the S-Band, the transmitter used two klystrons functioning in parallel, each with megawatt-level power. The MSR could search for targets from all directions, acquiring them at up to 300 miles (480 km) range.

One Safeguard site, intended to defend Minuteman ICBM missile silos near the Grand Forks AFB in North Dakota, was finally completed in October 1975, but the U.S. Congress withdrew all funding after it was operational but a single day. During the following decades, the U.S. Army and the U.S. Air Force developed a variety of large radar systems, but the long-serving BTL gave up military development work in the 1970s.

A modern radar developed by of the U.S. Navy that should be noted is the AN/SPY-1. First fielded in 1973, this S-Band, 6 MW system has gone through a number of variants and is a major component of the Aegis Combat System. An automatic detect-and-track system, it is computer controlled using four complementary three-dimensional passive electronically scanned array antennas to provide hemispherical coverage.

Radar signals, traveling with line-of-sight propagation, normally have a range to ground targets limited by the visible horizon, or less than about 10 miles (16 km). Airborne targets can be detected by ground-level radars at greater ranges, but, at best, several hundred miles. Since the beginning of radio, it had been known that signals of appropriate frequencies (3 to 30 MHz) could be "bounced" from the ionosphere and received at considerable distances. As long-range bombers and missiles came into being, there was a need to have radars give early warnings at great ranges. In the early 1950s, a team at the Naval Research Laboratory came up with the Over-the-Horizon (OTH) radar for this purpose.

To distinguish targets from other reflections, it was necessary to use a phase-Doppler system. Very sensitive receivers with low-noise amplifiers had to be developed. Since the signal going to the target and returning had a propagation loss proportional to the range raised to the fourth power, a powerful transmitter and large antennas were required. A digital computer with considerable capability (new at that time) was necessary for analyzing the data. In 1950, their first experimental system was able to detect rocket launches 600 miles (970 km) away at Cape Canaveral, and the cloud from a nuclear explosion in Nevada 1,700 miles (2,700 km) distant.

In the early 1970s, a joint American-British project, code named Cobra Mist, used a 10-MW OTH radar at Orfordness (the birthplace of British radar), England, in an attempt to detect aircraft and missile launchings over the Western USSR. Because of US-USSR ABM agreements, this was abandoned within two years. In the same time period, the Soviets were developing a similar system; this successfully detected a missile launch at 2,500 km (1,600 mi). By 1976, this had matured into an operational system named *Duga* ("Arc" in English), but known to western intelligence as Steel Yard and called Woodpecker by radio amateurs and others who suffered from its interference – the transmitter was estimated to have a power of 10 MW. Australia, Canada, and France also developed OTH radar systems.

With the advent of satellites with early-warning capabilities, the military lost most of its interest in OTH radars. However, in recent years, this technology has been reactivated for detecting and tracking ocean shipping in applications such as maritime reconnaissance and drug enforcement.

Systems using an alternate technology have also been developed for over-the-horizon detection. Due to diffraction, electromagnetic surface waves are scattered to the rear of objects, and these signals can be detected in a direction opposite from high-powered transmissions. Called OTH-SW (SW for Surface Wave), Russia is using such a system to monitor the Sea of Japan, and Canada has a system for coastal surveillance.

Civil Aviation Radars

The post-war years saw the beginning of a revolutionary development in Air Traffic Control (ATC) – the introduction of radar. In 1946, the Civil Aeronautics Administration (CAA) unveiled an experimental radar-equipped tower for control of civil flights. By 1952, the CAA had begun its first routine use of radar for approach and departure control. Four years later, it placed a large order for long-range radars for use in *en route* ATC; these had the capability, at higher altitudes, to see aircraft within 200 nautical miles (370 km). In 1960, it became required for aircraft flying in certain areas to carry a radar transponder that identified the aircraft and helped improve radar performance. Since 1966, the responsible agency has been called the Federal Aviation Administration (FAA).

A Terminal Radar Approach Control (TRACON) is an ATC facility usually located within the vicinity of a large airport. In the US Air Force it is known as RAPCON (Radar Approach Control), and in the US Navy as a RATCF (Radar Air Traffic Control Facility). Typically, the TRACON controls aircraft within a 30 to 50 nautical mile (56 to 93 km) radius of the airport at an altitude between 10,000 and 15,000 feet (3,000 to 4,600 m). This uses one or more Airport Surveillance Radars (ASR-7, 8, & 9), sweeping the sky once every few seconds.

The Digital Airport Surveillance Radar (DASR) is a newer TRACON radar system, replacing the old analog systems with digital technology. The civilian nomenclature for this radar is the ASR-11, and AN/GPN-30 is used by the military. Two radar systems are included. The primary is an S-Band (~2.8 GHz) system with 25 kW pulse power. It provides 3-D tracking of target aircraft and also measures rainfall intensity. The secondary is a P-Band (~1.05 GHz) system with a peak-power of about 25 kW. It uses a transponder set to interrogate aircraft and receive operational data. The antennas for both systems rotate atop a tall tower.

Weather Radar

David Atlas

During World War II, military radar operators noticed noise in returned echoes due to weather elements like rain, snow, and sleet. Just after the war, military scientists returned to civilian life or continued in the Armed Forces and pursued their work in developing a use for those echoes. In the United States, David Atlas, for the Air Force group at first, and later for MIT, developed the first operational weather radars. In Canada, J.S. Marshall and R.H. Douglas formed the "Stormy Weather Group " in Montreal. Marshall and his doctoral student Walter Palmer are well known for their work on the drop size distribution in mid-latitude rain that led to understanding of the Z-R relation, which correlates a given radar reflectivity with the rate at which water is falling on the ground. In the United Kingdom, research continued to study the radar echo patterns and weather elements such as stratiform rain and convective clouds, and experiments were done to evaluate the potential of different wavelengths from 1 to 10 centimetres.

Between 1950 and 1980, reflectivity radars, which measure position and intensity of precipitation, were built by weather services around the world. In United States, the U.S. Weather Bureau, established in 1870 with the specific mission of to provide meteorological observations and giving notice of approaching storms, developed the WSR-1 (Weather Surveillance Radar-1), one of the first weather radars. This was a modified version of the AN/APS-2F radar, which the Weather Bureau acquired from the Navy. The WSR-1A, WSR-3, and WSR-4 were also variants of this radar. This was followed by the WSR-57 (Weather Surveillance Radar – 1957) was the first weather radar designed specifically for a national warning network. Using WWII technology based on vacuum tubes, it gave only coarse reflectivity data and no velocity information. Operating at 2.89 GHz (S-Band), it had a peak-power of 410 kW and a maximum range of about 580 mi (930 km). AN/FPS-41 was the military designation for the WSR-57.

The early meteorologists had to watch a cathode ray tube. During the 1970s, radars began to be standardized and organized into larger networks. The next significant change in the United States was the WSR-74 series, beginning operations in 1974. There were two types: the WSR-74S, for replacements and filling gaps in the WSR-57 national network, and the WSR-74C, primarily for local use. Both were transistor-based, and their primary technical difference was indicated by the letter, S band (better suited for long range) and C band, respectively. Until the 1990s, there were 128 of the WSR-57 and WSR-74 model radars were spread across that country.

The first devices to capture radar images were developed during the same period. The number of scanned angles was increased to get a three-dimensional view of the precipitation, so that horizontal cross-sections (CAPPI) and vertical ones could be performed. Studies of the organization of thunderstorms were then possible for the Alberta Hail Project in Canada and National Severe Storms Laboratory (NSSL) in the US in particular. The NSSL, created in 1964, began experimentation on dual polarization signals and on Doppler effect uses. In May 1973, a tornado devastated Union City, Oklahoma, just west of Oklahoma City. For the first time, a Dopplerized 10-cm wavelength radar from NSSL documented the entire life cycle of the tornado. The researchers discovered a mesoscale rotation in the cloud aloft before the tornado touched the ground : the tornadic vortex signature. NSSL's research helped convince the National Weather Service that Doppler radar was a crucial forecasting tool.

Between 1980 and 2000, weather radar networks became the norm in North America, Europe, Japan and other developed countries. Conventional radars were replaced by Doppler radars, which in addition to position and intensity of could track the relative velocity of the particles in the air. In the United States, the construction of a network consisting of 10 cm (4 in) wavelength radars, called NEXRAD or WSR-88D (Weather Service Radar 1988 Doppler), was started in 1988 following NSSL's research. In Canada, Environment Canada constructed the King City station, with a five centimeter research Doppler radar, by 1985; McGill University dopplerized its radar (J. S. Marshall Radar Observatory) in 1993. This led to a complete Canadian Doppler network between 1998 and 2004. France and other European countries switched to Doppler network by the end of the 1990s to early 2000s. Meanwhile, rapid advances in computer technology led to algorithms to detect signs of severe weather and a plethora of "products" for media outlets and researchers.

After 2000, research on dual polarization technology moved into operational use, increasing the amount of information available on precipitation type (e.g. rain vs. snow). "Dual polarization" means that microwave radiation which is polarized both horizontally and vertically (with respect to the ground) is emitted. Wide-scale deployment is expected by the end of the decade in some countries such as the United States, France, and Canada.

Since 2003, the U.S. National Oceanic and Atmospheric Administration has been experimenting with phased-array radar as a replacement for conventional parabolic antenna to provide more time resolution in atmospheric sounding. This would be very important in severe thunderstorms as their evolution can be better evaluated with more timely data.

Also in 2003, the National Science Foundation established the Engineering Research Center for Collaborative Adaptive Sensing of the Atmosphere, "CASA", a multidisciplinary, multi-university collaboration of engineers, computer scientists, meteorologists, and sociologists to conduct fundamental research, develop enabling technology, and deploy prototype engineering systems designed to augment existing radar systems by sampling the generally undersampled lower troposphere with inexpensive, fast scanning, dual polarization, mechanically scanned and phased array radars.

Mapping Radar

The plan position indicator, dating from the early days of radar and still the most common type of display, provides a map of the targets surrounding the radar location. If the radar antenna on an aircraft is aimed downward, a map of the terrain is generated, and the larger the antenna, the greater the image resolution. After centimeter radar came into being, downward-looking radars – the H2S (L-Band) and H2X (C-Band) – provided real-time maps used by the U.S. and Britain in bombing runs over Europe at night and through dense clouds.

In 1951, Carl Wiley led a team at Goodyear Aircraft Corporation (later Goodyear Aerospace) in developing a technique for greatly expanding and improving the resolution of radar-generated images. Called synthetic aperture radar (SAR), an ordinary-sized antenna fixed to the side of an aircraft is used with highly complex signal processing to give an image that would otherwise require a much larger, scanning antenna; thus, the name synthetic aperture. As each pulse is emitted, it is radiated over a lateral band onto the terrain. The return is spread in time, due to reflections from features at different distances. Motion of the vehicle along the flight path gives the horizontal increments. The amplitude and phase of returns are combined by the signal processor using Fourier transform techniques in forming the image. The overall technique is closely akin to optical holography.

Through the years, many variations of the SAR have been made with diversified applications resulting. In initial systems, the signal processing was too complex for on-board operation; the signals were recorded and processed later. Processors using optical techniques were then tried for generating real-time images, but advances in high-speed electronics now allow on-board processes for most applications. Early systems gave a resolution in tens of meters, but more recent airborne systems provide resolutions to about 10 cm. Current ultra-wideband systems have resolutions of a few millimeters.

Other Radars and Applications

There are many other post-war radar systems and applications. Only a few will be noted.

Radar Gun

The most widespread radar device today is undoubtedly the radar gun. This is a small, usually hand-held, Doppler radar that is used to detect the speed of objects, especially trucks and automobiles in regulating traffic, as well as pitched baseballs, runners, or other moving objects in sports. This device can also be used to measure the surface speed of water and continuously manufactured materials. A radar gun does not return information regarding the object's position; it uses the Doppler effect to measure the speed of a target. First developed in 1954, most radar guns operate with very low power in the X or Ku Bands. Some use infrared radiation or laser light; these are usually called LIDAR. A related technology for velocity measurements in flowing liquids or gasses is called laser Doppler velocimetry; this technology dates from the mid-1960s.

Impulse Radar

As pulsed radars were initially being developed, the use of very narrow pulses was examined. The pulse length governs the accuracy of distance measurement by radar – the shorter the pulse, the greater the precision. Also, for a given pulse repetition frequency (PRF), a shorter pulse results in a higher peak power. Harmonic analysis shows that the narrower the pulse, the wider the band of frequencies that contain the energy, leading to such systems also being called wide-band radars. In the early days, the electronics for generating and receiving these pulses was not available; thus, essentially no applications of this were initially made.

By the 1970s, advances in electronics led to renewed interest in what was often called short-pulse radar. With further advances, it became practical to generate pulses having a width on the same order as the period of the RF carrier (T = 1/f). This is now generally called impulse radar.

The first significant application of this technology was in ground-penetrating radar (GPR). Developed in the 1970s, GPR is now used for structural foundation analysis, archeological mapping, treasure hunting, unexploded ordnance identification, and other shallow investigations. This is possible because impulse radar can concisely locate the boundaries between the general media (the soil) and the desired target. The results, however, are non-unique and are highly dependent upon the skill of the operator and the subsequent interpretation of the data.

In dry or otherwise favorable soil and rock, penetration up to 300 feet (91 m) is often possible. For distance measurements at these short ranges, the transmitted pulse is usually only one radio-frequency cycle in duration; With a 100 MHz carrier and a PRF of 10 kHz (typical parameters), the pulse duration is only 10 ns (nanosecond). leading to the "impulse" designation. A variety of GPR systems are commercially available in back-pack and wheeled-cart versions with pulse-power up to a kilowatt.

With continued development of electronics, systems with pulse durations measured in picoseconds became possible. Applications are as varied as security and motion sensors, building stud-finders, collision-warning devices, and cardiac-dynamics monitors. Some of these devices are matchbox sized, including a long-life power source.

Radar Astronomy

As radar was being developed, astronomers considered its application in making observations of the Moon and other near-by extraterrestrial objects. In 1944, Zoltán Lajos Bay had this as a major objective as he developed a radar in Hungary. His radar telescope was taken away by the conquering Soviet army and had to be rebuilt, thus delaying the experiment. Under Project Diana conducted by the Army's Evans Signal Laboratory in New Jersey, a modified SCR-271 radar (the fixed-position version of the SCR-270) operating at 110 MHz with 3 kW peak-power, was used in receiving echoes from the Moon on January 10, 1946. Zoltán Bay accomplished this on the following February 6.

Radio astronomy also had its start following WWII, and many scientists involved in radar development then entered this field. A number of radio observatories were constructed during the following years; however, because of the additional cost and complexity of involving transmitters and associated receiving equipment, very few were dedicated to radar astronomy. In fact, essentially all major radar astronomy activities have been conducted as adjuncts to radio astronomy observatories.

The radio telescope at the Arecibo Observatory, opened in 1963, is the largest in the world. Owned by the U.S. National Science Foundation and contractor operated, it is used primarily for radio astronomy, but equipment is available for radar astronomy. This includes transmitters operating at 47 MHz, 439 MHz, and 2.38 GHz, all with very-high pulse power. It has a 305-m (1,000-ft) primary reflector fixed in position; the secondary reflector is on tracks to allow precise pointing to different parts of the sky. Many significant scientific discoveries have been made using the Arecibo radar telescope, including mapping of surface roughness of Mars and observations of Saturns and its largest moon, Titan. In 1989, the observatory radar-imaged an asteroid for the first time in history.

Several spacecraft orbiting the Moon, Mercury, Venus, Mars, and Saturn have carried radars for surface mapping; a ground-penetration radar was carried on the Mars Express mission. Radar systems on a number of aircraft and orbiting spacecraft have mapped the entire Earth for various purposes; on the Shuttle Radar Topography Mission, the entire planet was mapped at a 30-m resolution.

The Jodrell Bank Observatory, an operation of the University of Manchester in Britain, was originally started by Bernard Lovell to be a radar astronomy facility. It initially used a war-surplus GL-II radar system operating at 71 MHz (4.2 m). The first observations were of ionized trails in the Geminids meteor shower during December 1945. While the facility soon evolved to become the third largest radio observatory in the world, some radar astronomy continued. The largest (250-ft or 76-m in diameter) of their three fully steerable radio telescopes became operational just in time to radar track *Sputnik 1*, the first artificial satellite, in October 1957.

Radar Engineering Details

Radar engineering details are technical details pertaining to the components of a radar and their ability to detect the return energy from moving scatterers — determining an object's position or obstruction in the environment. This includes field of view in terms of solid angle and maximum unambiguous range and velocity, as well as angular, range and velocity resolution. Radar sensors are classified by application, architecture, radar mode, platform, and propagation window.

Applications of radar include Autonomous cruise control system, autonomous landing guidance, radar altimeter, air traffic management, early-warning radar, fire-control radar, forward warning collision sensing, ground penetrating radar, surveillance, and weather forecasting.

Architecture Choice

The angle of a target is detected by scanning the field of view with a highly directive beam. This is done electronically, with a phased array antenna, or mechanically by rotating a physical antenna. The emitter and the receiver can be in the same place, as with the monostatic radars, or be separated as in the bistatic radars. Finally, the radar wave emitted can be continuous or pulsed. The choice of the architecture depends on the sensors to be used.

Scanning Antenna

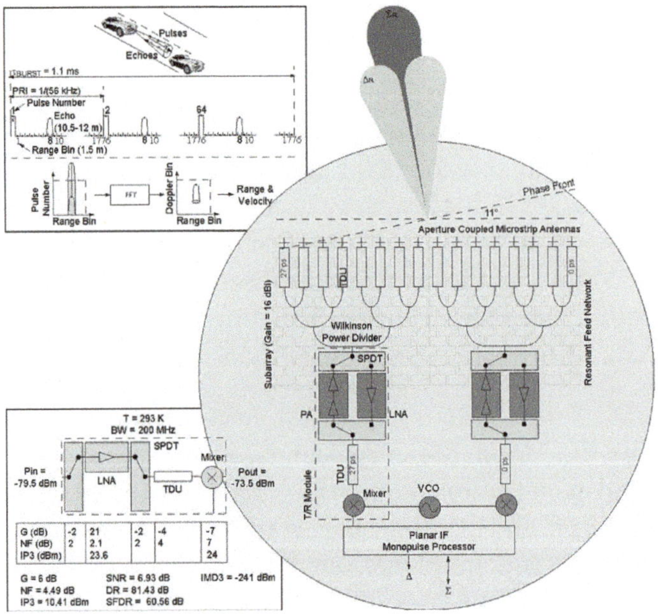

A passive electronically scanned array with a monopulse feed network.

An electronically scanned array (ESA), or a phased array, offers advantages over mechanically scanned antennas such as instantaneous beam scanning, the availability of multiple concurrent agile beams, and concurrently operating radar modes. Figures of merit of an ESA are the bandwidth, the effective isotropically radiated power (EIRP) and the G_R/T quotient, the field of view. EIRP is the product of the transmit gain, G_T, and the transmit power, P_T. G_R/T is the quotient of the receive gain and the antenna noise temperature. A high EIRP and G_R/T are a prerequisite for long-range detection. Design choices are:

- Active versus passive: In an active electronically scanned array (AESA), each antenna is connected to a T/R module featuring solid state power amplification (SSPA). An AESA has distributed power amplification and offers high performance and reliability, but is expensive. In a passive electronically scanned array, the array is connected to a single T/R module featuring vacuum electronics devices (VED). A PESA has centralized power amplification and offers cost savings, but requires low-loss phase shifters.

- Aperture: The Antenna aperture of a radar sensor is real or synthetic. Real-beam radar sensors allow for real-time target sensing. Synthetic aperture radar (SAR) allow for an angular resolution beyond real beamwidth by moving the aperture over the target, and adding the echoes coherently.

- Architecture: The field of view is scanned with a highly directive frequency-orthogonal (slotted waveguide), spatially orthogonal (switched beamforming networks), or time-orthogonal beams. In case of time-orthogonal scanning, the beam of an ESA is scanned preferably by applying a progressive time delay, $\Delta\tau$, constant over frequency, instead of by applying a progressive phase shift, constant over frequency. Usage of true-time-delay (TTD) phase shifters avoids beam squinting with frequency. The scanning angle, θ, is expressed as a function of the phase shift progression, β, which is a function of the frequency and the progressive time delay, $\Delta\tau$, which is invariant with frequency:

$$kd\cos\theta = \beta(f) = 2\pi\frac{c}{\lambda_0}\Delta\tau$$

$$\theta = \arccos\left(\frac{c}{d}\Delta\tau\right)$$

Note that θ is not a function of frequency. A constant phase shift over frequency has important applications as well, albeit in wideband pattern synthesis. For example, the generation of wideband monopulse Σ/Δ receive patterns depends on a feed network which combines two subarrays using a wideband hybrid coupler.

- Beam forming: The beam is formed in the digital (digital beamforming (DBF)), intermediate frequency (IF), optical, or radio frequency (RF) domain.
- Construction: An electronically scanned array is a brick, stick, tile, or tray construction. Brick and tray refers to a construction approach in which the RF circuitry is integrated perpendicular to the array plane. Tile, on the other hand, refers to a construction approach in which the RF circuitry is integrated on substrates parallel to the array plane. Stick refers to a construction approach in which the RF circuitry is connected to a line array in the array plane.
- Feed Network: The feed network is constrained (corporate, series) or space-fed.
- Grid: The grid is periodic (rectangular, triangular) or aperiodic (thinned).
- Polarization (antenna): The polarization of ground-based radar sensors is vertical, in order to reduce multipath (Brewster angle). Radar sensors can also be polarimetric for all-weather applications.

FMCW Versus Pulse-Doppler

The range and velocity of a target are detected through pulse delay ranging and the Doppler effect (pulse-Doppler), or through the frequency modulation (FM) ranging and range differentiation. The range resolution is limited by the instantaneous signal bandwidth of the radar sensor in both pulse-Doppler and frequency modulated continuous wave (FMCW) radars. Monostatic monopulse-Doppler radar sensors offer advantages over FMCW radars, such as:

- Half-duplex: Pulse-Doppler radar sensors are half-duplex, while FMCW radar sensors are full-duplex. Hence, pulse-Doppler provide higher isolation between transmitter and receiver, increasing the receiver's dynamic range (DR) and the range detection considerably. In addition, an antenna or an array can be time-shared between transmitter and receiver of the T/R module, whereas FMCW radars require two antennas or arrays, one for transmit and one for receive. A drawback of half-duplex operation is the existence of a blind zone in the immediate vicinity of the radar sensor. Pulse-Doppler radar sensors are therefore more suited for long-range detection, while FMCW radar sensors are more suited for short-range detection.

- Monopulse: A monopulse feed network, as shown in the fig. increases the angular accuracy to a fraction of the beamwidth by comparing echoes, which originate from a single radiated pulse and which are received in two or more concurrent and spatially orthogonal beams.
- Pulse compression: Pulse compression derelates the pulse width and the instantaneous signal bandwidth, which are otherwise inversely related. The pulse width is related to the time-on-target, the signal to noise ratio (SNR) and the maximum range. The instantaneous signal bandwidth is related to the range resolution.
- Pulse-Doppler processing: Echoes originating from a radiated burst are transformed to the spectral domain using a discrete Fourier transform (DFT). In the spectral domain, stationary clutter can be removed because it has a Doppler frequency shift which is different from the Doppler frequency shift of the moving target. The range and velocity of a target can be estimated with increased SNR due to coherent integration of echoes.

Bistatic Versus Monostatic

Bistatic radars have a spatially dislocated transmitter and receiver. In this case sensor in the transmitting antenna report back to the system the angular position of the scanning beam while the energy detecting ones are with the other antenna. A time synchronisation is crucial in interpreting the data as the receiver antenna is not moving.

Monostatic radars have a spatially co-located transmitter and receiver. It this case, the emission has to be insulated from the reception sensors as the energy emitted is far greater than the returned one.

Platform

Radar clutter is platform-dependent. Examples of platforms are airborne, car-borne, ship-borne, space-borne, and ground-based platforms.

Propagation Window

The radar frequency is selected based on size and technology readiness level considerations. The radar frequency is also chosen in order to optimize the radar cross-section (RCS) of the envisioned target, which is frequency-dependent. Examples of progation windows are the 3 GHz (S), 10 GHz (X), 24 GHz (K), 35 GHz (Ka), 77 GHz (W), 94 GHz (W) propagation windows.

Radar Mode

Radar modes for point targets include search and track. Radar modes for distributed targets include ground mapping and imaging. The radar mode sets the radar waveform

Marine Radar

Marine Radar Display as part of an integrated bridge system

Typical marine radar antenna on the stern of a small boat. It rotates on a vertical axis, scanning 360° of azimuth about every 2 seconds. It radiates a narrow vertical fan-shaped beam of microwaves horizontally toward the horizon, perpendicular to the long axis of the antenna.

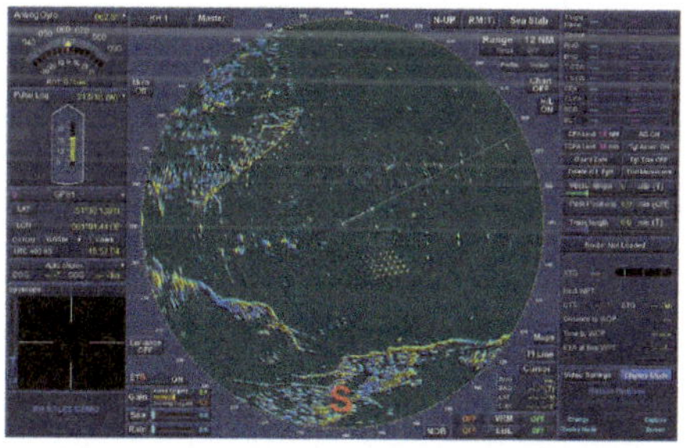

New Widescreen colour Marine Radar Display

Marine radars are X band or S band radars to provide bearing and distance of ships and land targets in vicinity from own ship (radar scanner) for collision avoidance and navigation at sea.

Radar is a vital component for safety at sea and near the shore. Captains need to be able to maneuver their ships within feet in the worst of conditions and to be able to navigate "blind". This means inside a dark room with no visibility they need to safely navigate their way through waters in the worst of weather. Radars are rarely used alone in a marine setting. In commercial ships, they are integrated into a full system of marine instruments including chartplotters, sonar, two-way radio communication devices, and emergency locators (SART).

The integration of these devices is very important as it becomes quite distracting to look at several different screens. Therefore, displays can often overlay charting, radar, sonar into a single system.

This gives the captain unprecedented instrumentation to maneuver the ship. With digital backbones, these devices have advanced greatly in the last years. For example, the newer ones have 3D displays that allow you to see above, below and all around the ship, including overlays of satellite imaging.

In port or in harbour, shore-based vessel traffic service radar systems are used to monitor and regulate ship movements in busy waters.

Collision Avoidance

As required by COLREGS, all ships shall maintain a proper radar lookout if it is available on board to obtain early warning of risk of collision. Radar plotting or the ARPA should be used to get the information of movement and the risk of collision (bearing, distance, CPA (closest point of approach), TCPA (time of closest point of approach) of other ships in vicinity.

Navigation

Marine radar systems can provide very useful radar navigation information for navigators onboard ships. The ship's position could be fixed by the bearing and distance information of a fixed, reliable target on the radar screen.

Radar controls

Marine radar has performance adjustment controls for brightness and contrast, also manual or automatic adjustment of gain, tuning, sea clutter and rain clutter suppression, and interference reduction. Other common controls consist of range scale, bearing cursor, fix/variable range marker or bearing/distance cursor.

Planar Array Radar

The planar array radar is a type of radar that uses a high-gain planar array antenna.

Operation

A fixed delay is established between horizontal arrays in the elevation plane. As the frequency is changed, the phase front across the aperture tends to tilt, with the result that the beam is moved in elevation. The differing frequencies cause each successive beam to be elevated slightly more than previous beams. A 27.5-degree elevation is scanned by the radar.

Advantages

- Each beam group has full transmitter peak power, full antenna gain and full antenna side-lobe performance.
- The use of frequency changes allows economical, simple and reliable inertialess elevation scanning.

Doppler Radar

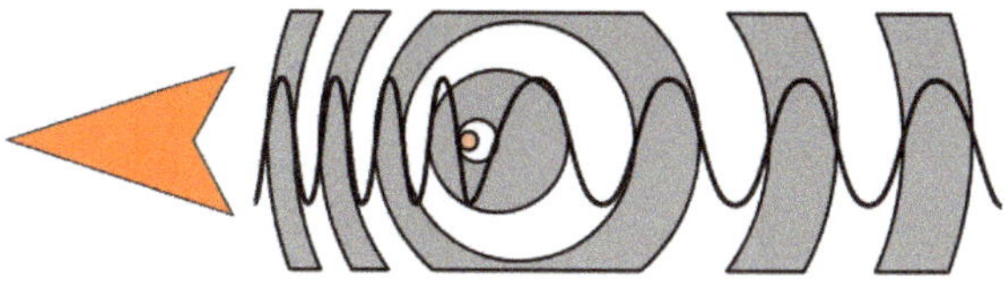

Doppler effect

A Doppler radar is a specialized radar that uses the Doppler effect to produce velocity data about objects at a distance. It does this by bouncing a microwave signal off a desired target and analyzing how the object's motion has altered the frequency of the returned signal. This variation gives direct and highly accurate measurements of the radial component of a target's velocity relative to the radar. Doppler radars are used in aviation, sounding satellites, meteorology, radar guns, radiology and healthcare (fall detection and risk assessment, nursing or clinic purpose), and bistatic radar (surface-to-air missiles).

Partly because of its common use by television meteorologists in on-air weather reporting, the specific term "*Doppler Radar*" has erroneously become popularly synonymous with the type of radar used in meteorology. Most modern weather radars use the pulse-Doppler technique to examine the motion of precipitation, but it is only a part of the processing of their data. So, while these radars use a highly specialized form of *Doppler radar*, the term is much broader in its meaning and its applications.

Concept

Doppler Effect

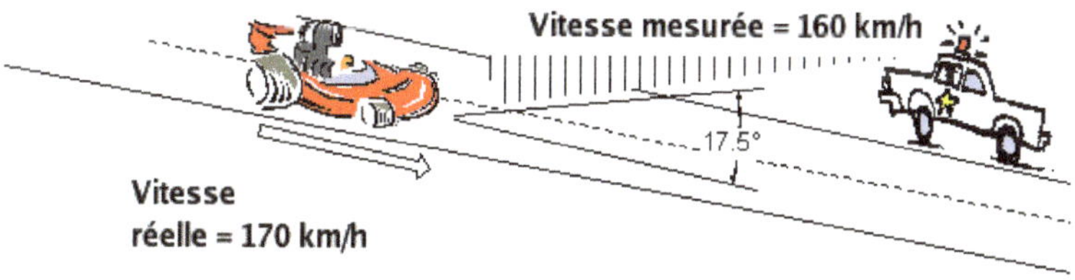

The emitted signal toward the car is reflected back with a variation of frequency that depend on the speed away/toward the radar (160 km/h). This is only a component of the real speed (170 km/h).

The Doppler effect (or Doppler shift), named after Austrian physicist Christian Doppler who proposed it in 1842, is the difference between the observed frequency and the emitted frequency of a wave for an observer moving relative to the source of the waves. It is commonly heard when a vehicle sounding a siren approaches, passes and recedes from an observer. The received frequency is higher (compared to the emitted frequency) during the approach, it is identical at the instant of passing by, and it is lower during the recession. This variation of frequency also depends on the direction the wave source is moving with respect to the observer; it is maximum when the source is moving directly toward or away from the observer and diminishes with increasing angle between the direction of motion and the direction of the waves, until when the source is moving at right angles to the observer, there is no shift.

Imagine a baseball pitcher throwing one ball every second to a catcher (a frequency of 1 ball per second). Assuming the balls travel at a constant velocity and the pitcher is stationary, the catcher catches one ball every second. However, if the pitcher is jogging towards the catcher, the catcher catches balls more frequently because the balls are less spaced out (the frequency increases). The inverse is true if the pitcher is moving away from the man. He catches balls less frequently because of the pitcher's backward motion (the frequency decreases). If the pitcher moves at an angle, but at the same speed, the frequency variation at which the receiver catches balls is less, as the distance between the two changes more slowly.

From the point of view of the pitcher, the frequency remains constant (whether he's throwing balls or transmitting microwaves). Since with electromagnetic radiation like microwaves frequency is inversely proportional to wavelength, the wavelength of the waves is also affected. Thus, the relative difference in velocity between a source and an observer is what gives rise to the doppler effect.

Frequency Variation

The formula for radar Doppler shift is the same as that for reflection of light by a moving mirror. There is no need to invoke Einstein's theory of special relativity, because all observations are made in the same frame of reference. The result derived with c as the speed of light and v as the target velocity gives the shifted frequency (f_r) as a function of the original frequency (f_t) :

$$f_r = f_t\left(\frac{1+v/c}{1-v/c}\right)$$

The "beat frequency", (Doppler frequency) (f_d), is thus:

$$f_d = f_r - f_t = 2v\frac{f_t}{(c-v)}$$

Since for most practical applications of radar, $v \ll c$, so $(c-v) \rightarrow c$. We can then write:

$$f_d \approx 2v\frac{f_t}{c}$$

Technology

U.S. Army soldier using a radar gun, an application of Doppler radar, to catch speeding violators.

There are four ways of producing the Doppler effect. Radars may be:

- Coherent pulsed (CP),
- Pulse-Doppler radar,
- Continuous wave (CW), or
- Frequency modulation (FM).

Doppler allows the use of narrow band receiver filters that reduce or eliminate signals from slow moving and stationary objects. This effectively eliminates false signals produced by trees, clouds, insects, birds, wind, and other environmental influences. Cheap hand held Doppler radar may produce erroneous measurements.

CW Doppler radar only provides a velocity output as the received signal from the target is compared in frequency with the original signal. Early Doppler radars included CW, but these quickly led to the development of frequency modulated continuous wave (FMCW) radar, which sweeps the transmitter frequency to encode and determine range.

With the advent of digital techniques, Pulse-Doppler radars (PD) became light enough for aircraft use, and Doppler processors for coherent pulse radars became more common. That provides Look-down/shoot-down capability. The advantage of combining Doppler processing with pulse radars is to provide accurate velocity information. This velocity is called range-rate. It describes the rate that a target moves toward or away from the radar. A target with no range-rate reflects a frequency near the transmitter frequency and cannot be detected. The classic zero doppler target is one which is on a heading that is tangential to the radar antenna beam. Basically, any target that is heading 90 degrees in relation to the antenna beam cannot be detected by its velocity (only by its conventional reflectivity).

In military airborne applications, the Doppler effect has 2 main advantages. Firstly, the radar is more robust against counter-measure. Return signals from weather, terrain, and countermeasures like chaff are filtered out before detection, which reduces computer and operator loading in hostile environments. Secondly, against a low altitude target, filtering on the radial speed is a very effective way to eliminate the ground clutter that always has a null speed. Low-flying military plane with countermeasure alert for hostile radar track acquisition can turn perpendicular to the hostile radar to nullify its Doppler frequency, which usually breaks the lock and drives the radar off by hiding against the ground return which is much larger.

History

Doppler radar tends to be lightweight because it eliminates heavy pulse hardware. The associated filtering removes stationary reflections while integrating signals over a longer time span, which improves range performance while reducing power. The military applied these advantages during the 1940s.

Continuous-broadcast, or FM, radar was developed during World War II for United States Navy aircraft, to support night combat operation. Most used the UHF spectrum and had a transmit Yagi antenna on the port wing and a receiver Yagi antenna on the starboard wing. This enabled bombers to fly an optimum speed when approaching ship targets, and let escort fighter aircraft train guns on enemy aircraft during night operation. These strategies were adapted to semi-active radar homing.

AN/APN-81 Doppler radar navigation system, mid-1950s

Modern Doppler systems are light enough for mobile ground surveillance associated with infantry and surface ships. These detect motion from vehicles and personnel for night and all weather combat operation. Modern police radar are a smaller, more portable version of these systems.

Early Doppler radar sets relied on large analog filters to achieve acceptable performance. Analog filters, waveguide, and amplifiers pick up vibration like microphones, so bulky vibration damping is required. That extra weight imposed unacceptable kinematic performance limitations that restricted aircraft use to night operation, heavy weather, and heavy jamming environments until the 1970s.

Digital fast Fourier transform (FFT) filtering became practical when modern microprocessors became available during the 1970s. This was immediately connected to coherent pulsed radars, where velocity information was extracted. This proved useful in both weather and air traffic control radars. The velocity information provided another input to the software tracker, and improved computer tracking. Because of the low pulse repetition frequency (PRF) of most coherent pulsed radars, which maximizes the coverage in range, the amount of Doppler processing is limited. The Doppler processor can only process velocities up to ±1/2 the PRF of the radar. This is not a problem for weather radars. Velocity information for aircraft cannot be extracted directly from low-PRF radar because sampling restricts measurements to about 75 miles per hour.

Specialized radars quickly were developed when digital techniques became lightweight and more affordable. Pulse-Doppler radars combine all the benefits of long range and high velocity capability. Pulse-Doppler radars use a medium to high PRF (on the order of 3 to 30 kHz), which allows for the detection of either high-speed targets or high-resolution velocity measurements. Normally it is one or the other; a radar designed for detecting targets from zero to Mach 2 does not have a high resolution in speed, while a radar designed for high-resolution velocity measurements does not have a wide range of speeds. Weather radars are high-resolution velocity radars, while air defense radars have a large range of velocity detection, but the accuracy in velocity is in the tens of knots.

Antenna designs for the CW and FM-CW started out as separate transmit and receive antennas before the advent of affordable microwave designs. In the late 1960s, traffic radars began being produced which used a single antenna. This was made possible by the use of circular polarization and a multi-port waveguide section operating at X band. By the late 1970s this changed to linear polarization and the use of ferrite circulators at both X and K bands. PD radars operate at too high a PRF to use a transmit-receive gas filled switch, and most use solid-state devices to protect the receiver low-noise amplifier when the transmitter is fired.

Doppler Navigation

Doppler navigation system in the National Electronics Museum

Wind Speed Correction

Doppler radars were used as a navigation aid for aircraft and spacecraft. By directly measuring the movement of the ground with the radar, and then comparing this to the airspeed returned from the aircraft instruments, the wind speed could be accurately determined for the first time. This value was then used for highly accurate dead reckoning. One early example of such a system was the Green Satin radar used in the English Electric Canberra. This system sent a pulsed signal at a very low repetition rate so it could use a single antenna to transmit and receive. An oscillator held the reference frequency for comparison to the received signal. In practice, the initial "fix" was taken using a radio navigation system, normally Gee, and the Green Satin then provided accurate long-distance navigation beyond Gee's 350-mile range. Similar systems were used in a number of aircraft of the era, and were combined with the main search radars of fighter designs by the 1960s.

Doppler navigation was in common commercial aviation use in the 1960s until it was largely superseded by inertial navigation systems. The equipment consisted of a transmitter/receiver unit, a processing unit and a gyro stabilised antenna platform. The antenna generated four beams and was rotated by a servo mechanism to align with the aircraft's track by equalising the Doppler shift from the left and right hand antennas. A synchro transmitted the platform angle to the flight deck, thus providing a measure of 'drift angle'. The ground speed was determined from the Doppler shift between the forward and aft facing beams. These were displayed on the flight deck on single instrument. Some aircraft had an additional 'Doppler Computer'. This was a mechanical device containing a steel ball rotated by a motor whose speed was controlled by the Doppler determined ground speed. The angle of this motor was controlled by the 'drift angle'. Two fixed wheels, one 'fore and aft' the other 'left to right' drove counters to output distance along track and across track difference. The aircraft's compass was integrated into the computer so that a desired track could be set between two waypoints on an over water great circle route. It may seem surprising to 21st. century readers, but it actually worked rather well and was great improvement over other 'dead reckoning' methods available at the time. It was generally backed up with position fixes from Loran, or as a last resort sextant and chronometer. It was possible to cross the Atlantic with an error of a couple of miles when in range of a couple of VORs or NDBs. It's major shortcoming in practice was the sea state, as a calm sea gave poor radar returns and hence unreliable Doppler measurements. But this was infrequent on the North Atlantic.

Locus-based Navigation

Location-based Doppler techniques were also used in the U.S. Navy's historical Transit satellite navigation system, with satellite transmitters and ground-based receivers, and are currently used in the civilian Argos system, which uses satellite receivers and ground-based transmitters. In these cases, the ground stations are either stationary or slow-moving, and the Doppler offset being measured is caused by the relative motion between the ground station and the fast-moving satellite. The combination of Doppler offset and reception time can be used to generate a locus of locations that would have the measured offset at that intersects the Earth's surface at that moment: by combining this with other loci from measurements at other times, the true location of the ground station can be determined accurately.

Doppler Effect

The Doppler effect causes a car engine or siren to sound higher in pitch when it is approaching than when it is receding. The pink circles represent sound waves.

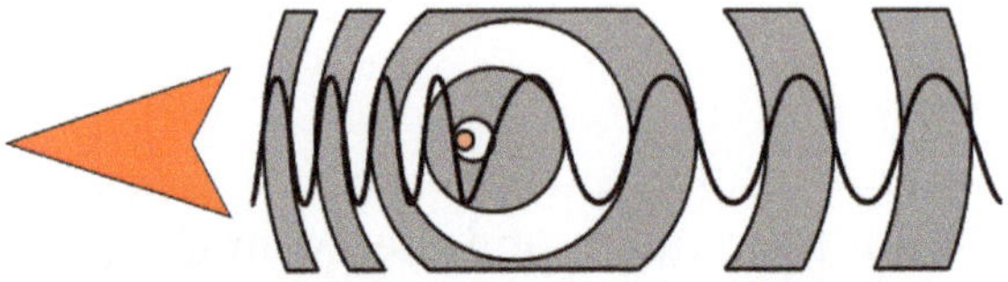

Change of wavelength caused by motion of the source.

Doppler effect of water flow around a swan

The Doppler effect (or the Doppler shift) is the change in frequency or wavelength of a wave (or other periodic event) for an observer moving relative to its source. It is named after the Austrian physicist Christian Doppler, who proposed it in 1842 in Prague. A common example of Doppler shift is the change of pitch heard when a vehicle sounding a siren or horn approaches, passes, and recedes from an observer. Compared to the emitted frequency, the received frequency is higher during the approach, identical at the instant of passing by, and lower during the recession.

When the source of the waves is moving towards the observer, each successive wave crest is emitted from a position closer to the observer than the previous wave. Therefore, each wave takes slightly less time to reach the observer than the previous wave. Hence, the time between the arrival of successive wave crests at the observer is reduced, causing an increase in the frequency. While they are travelling, the distance between successive wave fronts is reduced, so the waves "bunch together". Conversely, if the source of waves is moving away from the observer, each wave is emitted from a position farther from the observer than the previous wave, so the arrival time between successive waves is increased, reducing the frequency. The distance between successive wave fronts is then increased, so the waves "spread out".

For waves that propagate in a medium, such as sound waves, the velocity of the observer and of the source are relative to the medium in which the waves are transmitted. The total Doppler effect may therefore result from motion of the source, motion of the observer, or motion of the medium. Each of these effects are analyzed separately. For waves which do not require a medium, such as light or gravity in general relativity, only the relative difference in velocity between the observer and the source needs to be considered.

Developments

Doppler first proposed this effect in 1842 in his treatise "*Über das farbige Licht der Doppelsterne und einiger anderer Gestirne des Himmels*" (On the coloured light of the binary stars and some other stars of the heavens). The hypothesis was tested for sound waves by Buys Ballot in 1845. He confirmed that the sound's pitch was higher than the emitted frequency when the sound source approached him, and lower than the emitted frequency when the sound source receded from him. Hippolyte Fizeau discovered independently the same phenomenon on electromagnetic waves in 1848 (in France, the effect is sometimes called "effet Doppler-Fizeau" but that name was not adopted by the rest of the world as Fizeau's discovery was six years after Doppler's proposal). In Britain, John Scott Russell made an experimental study of the Doppler effect (1848).

General

In classical physics, where the speeds of source and the receiver relative to the medium are lower than the velocity of waves in the medium, the relationship between observed frequency f and emitted frequency f_0 is given by:

$$f = \left(\frac{c + v_r}{c + v_s}\right) f_0$$

where

c is the velocity of waves in the medium;

v_r is the velocity of the receiver relative to the medium; positive if the receiver is moving towards the source (and negative in the other direction);

v_s is the velocity of the source relative to the medium; positive if the source is moving away from the receiver (and negative in the other direction).

The frequency is decreased if either is moving away from the other.

The above formula assumes that the source is either directly approaching or receding from the observer. If the source approaches the observer at an angle (but still with a constant velocity), the observed frequency that is first heard is higher than the object's emitted frequency. Thereafter, there is a monotonic decrease in the observed frequency as it gets closer to the observer, through equality when it is coming from a direction perpendicular to the relative motion (and was emitted at the point of closest approach; but when the wave is received, the source and observer will no longer be at their closest), and a continued monotonic decrease as it recedes from the observer. When the observer is very close to the path of the object, the transition from high to low frequency is very abrupt. When the observer is far from the path of the object, the transition from high to low frequency is gradual.

If the speeds v_s and v_r are small compared to the speed of the wave, the relationship between observed frequency f and emitted frequency f_0 is approximately.

Observed frequency	Change in frequency
$f = \left(1 + \frac{\Delta v}{c}\right) f_0$	$\Delta f = \frac{\Delta v}{c} f_0$

where

$$\Delta f = f - f_0$$

$\Delta v = v_r - v_s$ is the velocity of the receiver relative to the source: it is positive when the source and the receiver are moving towards each other.

Analysis

To understand what happens, consider the following analogy. Someone throws one ball every second at a man. Assume that balls travel with constant velocity. If the thrower is stationary, the man will receive one ball every second. However, if the thrower is moving towards the man, he will receive balls more frequently because the balls will be less spaced out. The inverse is true if the thrower is moving away from the man. So it is actually the *wavelength* which is affected; as a consequence, the received frequency is also affected. It may also be said that the velocity of the wave remains constant whereas wavelength changes; hence frequency also changes.

With an observer stationary relative to the medium, if a moving source is emitting waves with an actual frequency f_0 (in this case, the wavelength is changed, the transmission velocity of the wave keeps constant -- note that the *transmission velocity* of the wave does not depend on the *velocity of the source*), then the observer detects waves with a frequency f given by

$$f = \left(\frac{c}{c + v_s}\right) f_0$$

A similar analysis for a moving *observer* and a stationary source (in this case, the wavelength keeps constant, but due to the motion, the rate at which the observer receives waves -- and hence the *transmission velocity* of the wave [with respect to the observer] -- is changed) yields the observed frequency:

$$f = \left(\frac{c + v_r}{c}\right) f_0$$

These can be generalized into the equation that was presented in the previous section.

$$f = \left(\frac{c + v_r}{c + v_s} \right) f_0$$

An interesting effect was predicted by Lord Rayleigh in his classic book on sound: if the source is moving toward the observer at twice the speed of sound, a musical piece emitted by that source would be heard in correct time and tune, but *backwards*. The Doppler effect with sound is only clearly heard with objects moving at high speed, as change in frequency of musical tone involves a speed of around 40 meters per second, and smaller changes in frequency can easily be confused by changes in the amplitude of the sounds from moving emitters. Neil A Downie has demonstrated how the Doppler effect can be made much more easily audible by using an ultrasonic (e.g. 40 kHz) emitter on the moving object. The observer then uses a heterodyne frequency converter, as used in many bat detectors, to listen to a band around 40 kHz. In this case, with the bat detector tuned to give frequency for the stationary emitter of 2000 Hz, the observer will perceive a frequency shift of a whole tone, 240 Hz, if the emitter travels at 2 meters per second.

Application

Sirens

The siren on a passing emergency vehicle will start out higher than its stationary pitch, slide down as it passes, and continue lower than its stationary pitch as it recedes from the observer. Astronomer John Dobson explained the effect thus:

> "The reason the siren slides is because it doesn't hit you."

In other words, if the siren approached the observer directly, the pitch would remain constant until the vehicle hit him, and then immediately jump to a new lower pitch. Because the vehicle passes by the observer, the radial velocity does not remain constant, but instead varies as a function of the angle between his line of sight and the siren's velocity:

$$v_{\text{radial}} = v_s \cdot \cos\theta$$

where θ is the angle between the object's forward velocity and the line of sight from the object to the observer.

Astronomy

The Doppler effect for electromagnetic waves such as light is of great use in astronomy and results in either a so-called redshift or blueshift. It has been used to measure the speed at which stars and galaxies are approaching or receding from us; that is, their radial velocities. This may be used to detect if an apparently single star is, in reality, a close binary, to measure the rotational speed of stars and galaxies, or to detect exoplanets. (Note that redshift is also used to measure the expansion of space, but that this is not truly a Doppler effect.) This redshift and blueshift happens on a very small scale, if an object is moving toward earth, there would not be a noticeable difference in visible light.

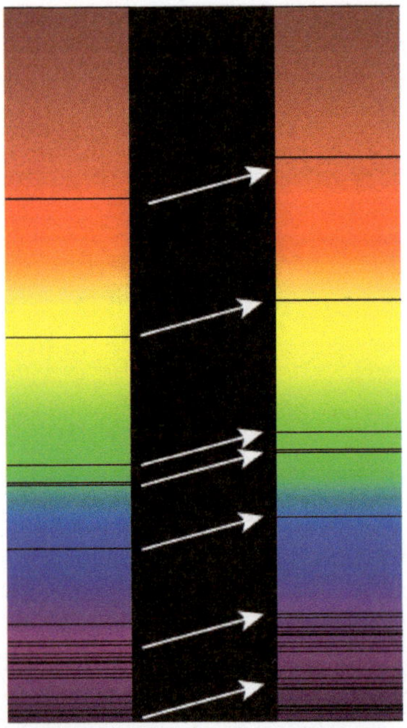

Redshift of spectral lines in the optical spectrum of a supercluster of distant galaxies (right), as compared to that of the Sun (left)

The use of the Doppler effect for light in astronomy depends on our knowledge that the spectra of stars are not homogeneous. They exhibit absorption lines at well defined frequencies that are correlated with the energies required to excite electrons in various elements from one level to another. The Doppler effect is recognizable in the fact that the absorption lines are not always at the frequencies that are obtained from the spectrum of a stationary light source. Since blue light has a higher frequency than red light, the spectral lines of an approaching astronomical light source exhibit a blueshift and those of a receding astronomical light source exhibit a redshift.

Among the nearby stars, the largest radial velocities with respect to the Sun are +308 km/s (BD-15°4041, also known as LHS 52, 81.7 light-years away) and -260 km/s (Woolley 9722, also known as Wolf 1106 and LHS 64, 78.2 light-years away). Positive radial velocity means the star is receding from the Sun, negative that it is approaching.

Radar

The Doppler effect is used in some types of radar, to measure the velocity of detected objects. A radar beam is fired at a moving target — e.g. a motor car, as police use radar to detect speeding motorists — as it approaches or recedes from the radar source. Each successive radar wave has to travel farther to reach the car, before being reflected and re-detected near the source. As each wave has to move farther, the gap between each wave increases, increasing the wavelength. In some situations, the radar beam is fired at the moving car as it approaches, in which case each successive wave travels a lesser distance, decreasing the wavelength. In either situation, calculations from the Doppler effect accurately determine the car's velocity. Moreover, the proximity fuze, developed during World War II, relies upon Doppler radar to detonate explosives at the correct time, height, distance, etc.

Because the doppler shift affects the wave incident upon the target as well as the wave reflected back to the radar, the change in frequency observed by a radar due to a target moving at relative velocity Δv is twice that from the same target emitting a wave:

$$\Delta f = \frac{2\Delta v}{c} f_0.$$

Medical Imaging and Blood Flow Measurement

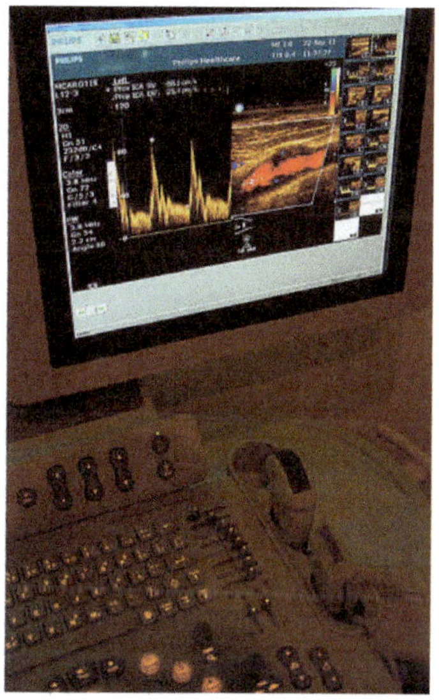

Colour flow ultrasonography (Doppler) of a carotid artery - scanner and screen

An echocardiogram can, within certain limits, produce an accurate assessment of the direction of blood flow and the velocity of blood and cardiac tissue at any arbitrary point using the Doppler effect. One of the limitations is that the ultrasound beam should be as parallel to the blood flow as possible. Velocity measurements allow assessment of cardiac valve areas and function, abnormal communications between the left and right side of the heart, leaking of blood through the valves (valvular regurgitation), and calculation of the cardiac output. Contrast-enhanced ultrasound using gas-filled microbubble contrast media can be used to improve velocity or other flow-related medical measurements.

Although "Doppler" has become synonymous with "velocity measurement" in medical imaging, in many cases it is not the frequency shift (Doppler shift) of the received signal that is measured, but the phase shift (*when* the received signal arrives).

Velocity measurements of blood flow are also used in other fields of medical ultrasonography, such as obstetric ultrasonography and neurology. Velocity measurement of blood flow in arteries and veins based on Doppler effect is an effective tool for diagnosis of vascular problems like stenosis.

Flow Measurement

Instruments such as the laser Doppler velocimeter (LDV), and acoustic Doppler velocimeter (ADV) have been developed to measure velocities in a fluid flow. The LDV emits a light beam and the ADV emits an ultrasonic acoustic burst, and measure the Doppler shift in wavelengths of reflections from particles moving with the flow. The actual flow is computed as a function of the water velocity and phase. This technique allows non-intrusive flow measurements, at high precision and high frequency.

Velocity Profile Measurement

Developed originally for velocity measurements in medical applications (blood flow), Ultrasonic Doppler Velocimetry (UDV) can measure in real time complete velocity profile in almost any liquids containing particles in suspension such as dust, gas bubbles, emulsions. Flows can be pulsating, oscillating, laminar or turbulent, stationary or transient. This technique is fully non-invasive.

Satellite Communication

Fast moving satellites can have a Doppler shift of dozens of kilohertz relative to a ground station. The speed, thus magnitude of Doppler effect, changes due to earth curvature. Dynamic Doppler compensation, where the frequency of a signal is changed multiple times during transmission, is used so the satellite receives a constant frequency signal.

Audio

The Leslie speaker, most commonly associated with and predominantly used with the famous Hammond organ, takes advantage of the Doppler effect by using an electric motor to rotate an acoustic horn around a loudspeaker, sending its sound in a circle. This results at the listener's ear in rapidly fluctuating frequencies of a keyboard note.

Vibration Measurement

A laser Doppler vibrometer (LDV) is a non-contact instrument for measuring vibration. The laser beam from the LDV is directed at the surface of interest, and the vibration amplitude and frequency are extracted from the Doppler shift of the laser beam frequency due to the motion of the surface.

Developmental Biology

During the segmentation of vertebrate embryos, waves of gene expression sweep across the presomitic mesoderm, the tissue from which the precursors of the vertebrae (somites) are formed. A new somite is formed upon arrival of a wave at the anterior end of the presomitic mesoderm. In zebrafish, it has been shown that the shortening of the presomitic mesoderm during segmentation leads to a Doppler effect as the anterior end of the tissue moves into the waves. This Doppler effect contributes to the period of segmentation.

Inverse Doppler Effect

Since 1968 scientists such as Victor Veselago have speculated about the possibility of an inverse Doppler effect. The experiment that claimed to have detected this effect was conducted by Nigel Seddon and Trevor Bearpark in Bristol, United Kingdom in 2003.

Researchers from many Universities like Swinburne University of Technology and the University of Shanghai for Science and Technology showed that this effect can be observed in optical frequencies as well. This was made possible by growing a photonic crystal and projecting a laser beam into the crystal. This made the crystal act like a super prism and the inverse Doppler effect could be observed.

Wave Radar

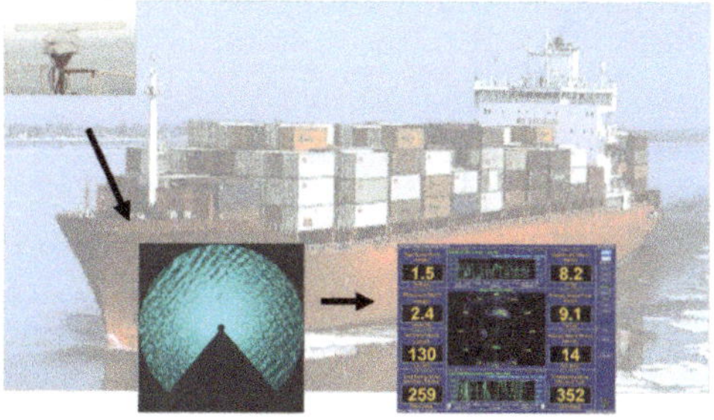

Measuring ocean waves by use of marine radars.

Wind waves can be measured by several radar remote sensing techniques. Several instruments based on a variety of different concepts and techniques are available to the user and these are all often called wave radars. gives a brief description of the most common ground-based radar remote sensing techniques.

Instruments based on radar remote sensing techniques have become of particular interest in applications where it is important to avoid direct contact with the water surface and avoid structural interference. A typical case is wave measurements from an offshore platform in deep waters with the presence of high currents making the mooring of a wave buoy enormously difficult. Another interesting case is a ship under way where having instruments in the sea is highly impractical and interference from the ship's hull must be avoided.

Radar Remote Sensing

Terms and Definitions

Basically there are two different *classes* of radar remote sensors for ocean waves.

- Direct sensor measures directly some relevant parameter of the wave system (like surface elevation or water particle velocity).
- Indirect sensors observe the surface waves via the interaction with some other physical process as for example the radar cross section of the sea surface.

Microwave radars may be used in two different *modes*;

- The near vertical mode. The radar echo is generated by specular reflections from the sea surface.
- The low grazing angle mode. The radar echo is generated by Bragg scattering, hence wind generated surface ripple (capillary waves) must be present. The backscattered signal will be

modulated by the large surface gravity waves and the gravity wave information is derived from the modulation of the backscattered signal. An excellent presentation of the theories of microwave remote sensing of the sea surface is given by Plant and Shuler (1980).

The radar footprint (the radial and azimuthal extent of the surface area to be illuminated by the radar) must be small in comparison with all ocean wavelength of interest. The radar spatial resolution is determined by the bandwidth of the radar signal and the beamwidth of the radar antenna.

The beam of a microwave antenna is dispersive, consequently the resolution becomes a function of range. The beam of an IR radar (laser) is non dispersive, the radar footprint is therefore independent of range.

HF radars utilize the Bragg scattering mechanism and do always operate at very low grazing angles. Due to the low frequency of operation the radar waves are backscattered directly from the gravity waves and surface ripple need not be present.

Radar transceivers may be coherent or non-coherent. Coherent radars measure Doppler-modulation as well as amplitude modulation, while non-coherent radars only measure amplitude modulation. Consequently, a non-coherent radar echo contains less information about the sea surface properties. Examples of non-coherent radars are conventional marine navigation radars.

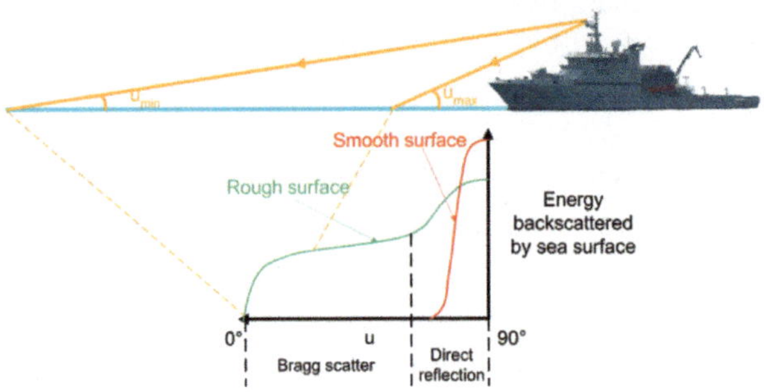

Energy backscattered from sea surface as a function of angle.

The radar transmitter waveform may be either unmodulated continuous wave, modulated or pulsed. An unmodulated continuous wave radar has no range resolution, but can resolve targets on the basis of different velocity, while a modulated or pulsed radar can resolve echoes from different ranges. The radar waveform plays a very important role in radar theory (Plant and Shuler, 1980).

The Wave Radar Performance is Highly Dependent on

- Mode of operation or measurement geometry (vertical or grazing)
- Class of system (direct or indirect)
- Frequency of operation
- Radar waveform (unmodulated CW or modulated/pulsed)
- Type of transceiver (coherent or non-coherent)
- Radar antenna properties

Remote Sensing Techniques

An excellent survey of different radar techniques for remote sensing of waves is given by Tucker (1991).

Microwave Range Finders

Microwave range finders also operate in vertical mode at GHz frequencies and is not as affected by fog and water spray as the laser altimeter. A continuous wave frequency modulated (CWFM) or pulsed radar waveform is normally used to provide range resolution. The beam is dispersive, hence the size of the footprint increases linearly with range.

One example of a microwave range finder is the Miros SM-094 which is designed for wave and water level (and tide) measurements. This sensor is applied as air gap (bridge clearance) sensor in NOAAs PORTS system. Another example is the WaveRadar REX which is a derivative of a Rosemount tank radar.

Digitized sea clutter image.

An array of three vertical radars in a triangular configuration can be used to measure a directional wave spectrum. Algorithms and signal processing software similar to what is used in the processing of heave, pitch, roll buoys. A commercial system called "Directional WaveGuide" is available from the Dutch companies Enraf and Radac.

Marine Navigation Radars

Marine navigation radars (X band) provide sea clutter images which contain a pattern resembling a sea wave pattern. By digitizing the radar video signal it can be processed by a digital computer. Sea surface parameters may be calculated on the basis of these digitized images. The marine navigation radar operates in low grazing angle mode and wind generated surface ripple must be present. The marine navigation radar is non-coherent and is a typical example of an indirect wave sensor, because there is no direct relation between wave height and radar back-scatter modulation amplitude. An empirical method of wave spectrum scaling is normally employed. Marine navigation radar based wave sensors are excellent tools for wave direction measurements. A marine navigation radar may also be a tool for surface current measurements. Point measurements of the current vector as well as current maps up to a distance of a few km can be provided (Gangeskar, 2002). Miros WAVEX has its main area of application as directional wave measurements from moving ships. Another example of a marine radar based system is OceanWaves WaMoS II.

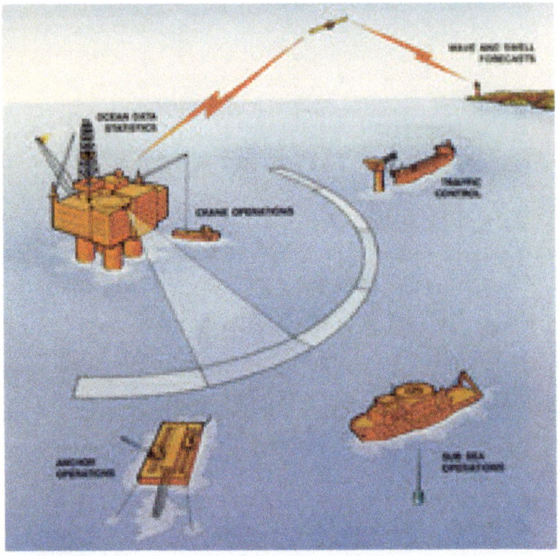

Measurement geometry of pulsed Doppler wave and current radar.

The Range Gated Pulsed Doppler Microwave Radar

The range gated pulsed Doppler microwave radar operates in low grazing angle mode. By using several antennas it may be used as a directional wave sensor, basically measuring the directional spectrum of the horizontal water particle velocity. The velocity spectrum is directly related to the wave height spectrum by a mathematical model based on linear wave theory and accurate measurements of the wave spectrum can be provided under most conditions. As measurements are taken at a distance from the platform on which it is mounted, the wave field is to a small degree disturbed by interference from the platform structure.

Miros Wave and current radar is the only available wave sensor based on the range gated pulsed Doppler radar technique. This radar also uses the dual frequency technique to perform point measurements of the surface current vector.

The Dual Frequency Microwave Radar

The dual frequency microwave radar transmits two microwave frequencies simultaneously. The frequency separation is chosen to give a "spatial beat" length which is in the range of the water waves of interest. The dual frequency radar may be considered a microwave equivalent of the high frequency (HF) radar. The dual frequency radar is suitable for the measurement of surface current. As far as wave measurements are concerned, the back-scatter processes are too complicated (and not well understood) to allow useful measurement accuracy to be attained.

The HF Radar

The HF radar CODAR SeaSonde and Helzel WERA are well established as a powerful tool for sea current measurements up to a range of 300 km. It operates in the HF and low VHF frequencies band corresponding to a radar wavelength in the range of 10 to 300m. The Doppler shift of the first order Bragg lines of the radar echo is used to derive sea current estimates in very much the same way as for the dual frequency microwave radar. Two radar installations are normally required, looking at the same patch of the sea surface from different angles. The latest generation of shore-based ocean radar can reach more than 200 km for ocean current mapping and more than 100 km for wave measurements Helzel WERA. For all ocean radars, the accuracy in range is excellent. With shorter ranges, the range resolution gets finer. The angular resolution and accuracy depends on the used antenna array configuration and applied algorithms (direction finding or beam forming). The WERA system provides the option to use both techniques; the compact version with direction finding or the array type antenna system with beam forming methods.

Specialized X-Band - FutureWaves™

The FutureWaves™ technology was originally developed as an Environmental Ship and Motion Forecasting (ESMF) system for the Navy's ONR (Office of Naval Research) by General Dynamics' Applied Physical Sciences Corporation. The technology was adapted to be released in the commercial market and made its first public appearance at the 2017 Offshore Technology Conference in Houston Texas.

FutureWaves™ differs from existing wave forecasting systems by employing a customized wave sensing radar capable of measuring backscatter Doppler out to ranges of approximately 5 km. The radar antenna is vertically polarized to enhance the sea-surface backscatter signal. Second, FutureWaves™ uses an innovative radar signal processing scheme that addresses the inherently noisy backscatter signals through a mathematical process termed Least Squares Inversion. This approach applies a highly over-determined filter to the radar data, and intelligently rejects radar scans that do not observe incoming waves. The result is an accurate representation of the propagating incident wave field that will force ship motions over a 2-3 minute window. The wave processing algorithms also enable real-time calculation of wave field two-dimensional power spectra and significant wave height similar to that provided by a wave buoy.

FutureWaves™ also uses an innovative vessel motion prediction process that relies on a pre-calculated forceresponse database. Dynamic motional degrees of freedom are then represented as a lumped mechanical system whose future motions are predicted by numerically solving a multi-degree-of-freedom, forced, coupled differential equation with initial inertial state provided by vessel motion sensor outputs. The time-domain solution employed in FutureWaves™ allows for nonlinear forcing mechanisms, such as quadratic roll damping and roll control systems, to be captured in the forecasting.

Finally, FutureWaves™ uses the Gravity open architecture middleware solution to integrate the sensor feeds, processing subroutines and user displays. This open architecture approach will allow FutureWaves™ users to implement customized operator displays along with physics based models of specific vessels and machinery (e.g. cranes) into the system.

Coastal Ocean Dynamics Applications Radar

Coastal ocean dynamics applications radar (CODAR) describes a type of portable, land-based, High Frequency (HF) radar developed between 1973 and 1983 at NOAA's Wave Propagation Laboratory in Boulder, Colorado. CODAR is a noninvasive system that permits to measure and map near-surface ocean currents in coastal waters. It is transportable and offers output ocean current maps on site in near real time. Moreover, using CODAR it is possible to measure waves heights and it provides an indirect estimate of local wind direction.

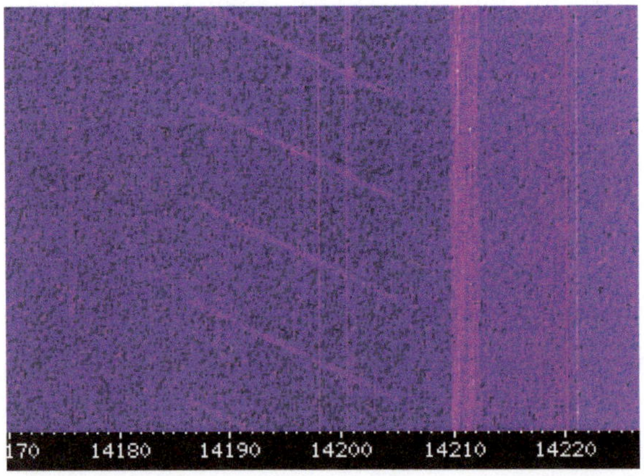

A CODAR source, detected in Atlanta, Georgia, on 14195 kHz, as seen from a software-defined radio's waterfall display. The sweeping, diagonal line is the CODAR signal.

Equipment

CODAR utilizes a compact antenna system that consists of crossed loops and a whip for receiving and a whip for transmitting radio pulses. The system can be transported by vehicle and it is capable of being operated from a portable power supply (for modern instrumentation it is recommended a minimum capacity of 1050 Watts. CODAR is capable of operating in virtually all weather conditions (it is tolerable a temperature from 0 °F (-18 °C) to 90 °F (32 °C) and the relatively small dimensions of the antenna system consent CODAR deployment even in highly populated and rocky coastal areas. However, as the signal is rapidly attenuated by land, the antenna has to be mounted as close to the surface water as possible.

The Operating Frequency Range of modern equipments goes from 3 to 50 MHz and it can be programmed for unattended operation for periods of up to two weeks.

The main equipment is cabled to the electronic segment, that is housed near-by in a sheltered environment and contains the system hardware, where information is stored. A minicomputer controls the radar and processes the signals and the operator can communicate with the system through a portable keyboard terminal.

The raw spectral data can be processed on-line to obtain real-time outputs and the final data products can be displayed on a graphics terminal or outputted to a hardcopy plotter. Otherwise, off-line processing at a later date can be accomplished.

Applications

Measuring surface currents is the primary result obtained by using CODAR. Ranges achieved and resolution vary with environmental conditions and antenna placement. In general, however, in their long-range mode, modern CODAR can measure out to 100–200 km offshore with a resolution of 3–12 km. By increasing the frequency, resolutions as fine as 200–500 m can be obtained, but the observation range is shortened (15–20 km).

However, the actual range can be limited by radio interferences, high-ocean states and ground conditions in the vicinity of the antennas. Wet and moist sandy soils enhance the ground wave propagation, whereas dry and rocky grounds reduce signal strengths.

A single CODAR system can measure only the component of surface current travelling toward or away from the radar so, to determinate the total surface current vectors, it is necessary to use at least a two-systems set up. A series of CODAR sites can be employed to obtain regional coverage. In a multiple radar configuration, spacing between two radar systems should be approximately 15 to 40 km for long-range open ocean mode and 8 to 20 km for "higher frequency, higher resolution, shorter range" mode.

Typically, CODAR data are averaged over one hour to reduce the noisiness of the sea echo. Therefore, current maps can be produced every hour. This period can be reduced to approximately 20 minutes, however the data may become noisy over such a short period.

CODAR's measurements are interesting for both military and civil purposes. Main applications are for example coastal engineering and public safety projects, planning of navigational seaways, mitigation of ocean pollution, search and rescue operations, oil-spill mitigation in real time and larval population connectivity assessment. Also, data obtained by using CODAR are used as inputs for global resource monitoring and weather forecasting models and are particularly helpful for tidal and storm-surge measurements. Moreover, from the measurements can be extracted the direction of propagation of wave energy and the period of the most energetic waves, which are important data for many practical applications in design and operation of coastal and offshore structures.

Theory of Operation

CODAR operates using sky transmission of waves in the high frequency (HF) band (3–30 MHz), as electromagnetic waves in this band have wavelengths that are commensurate with wind-driven gravity waves on the ocean surface. According as the customer necessities, it can be used in single or multi-frequency mode. As the ocean has a rough surface, when a high frequency signal reaches the ocean surface, a portion of the incident energy is scattered back towards the source and the receiver measures the reflected signal. This backscattering (or reflection) produces an energy spectrum at the receiver, even if the energy source is single-frequency, because of the shape and motion of the sea surface. Interpreting the spectral returns for various transmit frequencies is the key to extracting information about the ocean and, specifically, to measure surface currents.

As a consequence of Bragg's Scattering Law; the strongest received return comes from ocean waves traveling directly toward or away from the radar source and whose physical wavelength is exactly one-half as long as the transmitted radar wave. The return signal is processed and its spectral analysis provides the sea-echo Doppler spectrum, where two dominant peaks at different frequencies can be recognized.

Displacement of these peaks away from their known frequencies is called the "echo Doppler shift" and allows one to assess the radial velocity of a surface current. That is; the scatter velocity along the line between the hit surface and the radar. In fact, the magnitude of this component of the velocity is proportional to the degree of signal-shift. Therefore, CODAR measures the Doppler-induced frequency-shift (along with the distance from the radar to the sector and directional angle) to provide an estimate of the radial-component of wave-speed in the sector of sea surface of interest.

Measuring Surface Currents

In order to measure currents, the CODAR equipment computes three components:

- the velocity of incoming waves in the radial direction
- the distance from the radar equipment to the evaluated ocean sector
- the angle the waves are traveling relatively to the CODAR station

Computation of the Radial Velocity of Currents

The signal sent from the CODAR antenna has a known frequency and it moves at the speed of light. Therefore, the wavelength of the signal is known (wavelength = speed of light / frequency). Exploiting the Bragg's Law, CODAR maximizes the scattered HF signal, given that the resonance will only occur for the given wavelength:

$$\lambda\_s = \lambda\_t / (2 * \cos(\varphi))$$

where λ_s is the wavelength of surface ocean wave, λ_t is the wavelength of transmitted signal and φ is the angle of incidence between the signal and the ocean surface.

As the CODAR antennas are usually placed at sea level, the angle of incidence theta can be assumed to be zero. Therefore, the equation reduces to:

$$\lambda\_s = \lambda\_t / 2$$

This means that when the emitted signal hits waves with wavelength equal to one-half of the transmitted signal, the signal that is scattered back to the antenna will be in phase. Therefore, these waves will produce a scattered signal "stronger" and thus easily identifiable, which is measured by the CODAR system. Thus, the current speed is extracted by determining the Doppler Shift of the waves.

However, the above equations represent a simplified model, as they assume that the reflecting waves are not moving. This is of course untrue and, because of the motion, the frequency of the scattered signal (and therefore its wavelength) is not the same as the one of the transmitted signal. In fact, "waves moving toward the receiver increase the return frequency, while waves moving away decrease the return frequency".

Then a further Doppler shift (Δf) is observed and, by measuring it, it is possible to determine the radial velocity v_s component of the surface current by using the Doppler formula:

$$\Delta f = v\_s / \lambda\_s$$

Computation of the Distance to Target

The range to target is calculated starting from the time delay, which is obtained by subtracting the return signal time from the transmitted signal one.

Computation of the Angular Direction to Target

CODAR is a "direction finding system". The signal is received by two loop antennas and a monopole. Whereas the signal the monopole receives does not vary with the direction of the incoming signal, the signal received by the two loop antennas (positioned at a 90° angle) does vary with direction. This information permits a software to determine the direction of the signal.

Once calculated the radial velocity of currents, distance to target, and the angular direction to target it is possible to determine the current vector and to construct current vector maps. In fact, for the area in which vector data from two CODAR sites overlap, it is possible to calculate the velocity and direction of the current and comparisons with surface drifters and error analysis made in 1979 indicate that CODAR measures surface currents with at least 10 cm/s accuracy. In 2010, retailers of modern CODAR equipments guarantee an accuracy typically < 7 cm/s of the total current velocity and 1–2 cm of the tidal component, in normal environment condition. However, the accuracy of the system depends on several factors, such as signal-to-noise ratios, geometry and pointing errors.

Limitations

There are some limitations inherent to the system that do not permit certain applications. Here are presented the main practical limitations:

- CODAR does not permit to measure currents and waves in the area closer than 2 km from its location. This uncovered area results from the fact that the receiver is turned off during the pulse transmission. In this lapse of time, any signal backscattered is lost.
- CODAR's resolution cell size is generally bigger than 5 km^2. This does not permit to use this system for most of the inlets and harbor entrances.

As discussed before, for a given look angle, a single CODAR station permits to detect only the component of flow traveling toward or away from its location. Radial currents from two or more sites should be combined to obtain vector surface current estimates. Moreover, when using two CODAR stations the so-called "baseline problem" can affect the measurement. This occurs when both the instruments measure the same component of velocity. To avoid this problem and resolve properly the current vector, generally two radials must have an angle included by 30° and 150°.

Radar Signal Characteristics

A radar system uses a radio frequency electromagnetic signal reflected from a target to determine information about that target. In any radar system, the signal transmitted and received will exhibit many of the characteristics described below.

The Radar Signal in the Time Domain

The diagram below shows the characteristics of the transmitted signal in the time domain. Note that in this and in all the diagrams within this article, the x axis is exaggerated to make the explanation clearer.

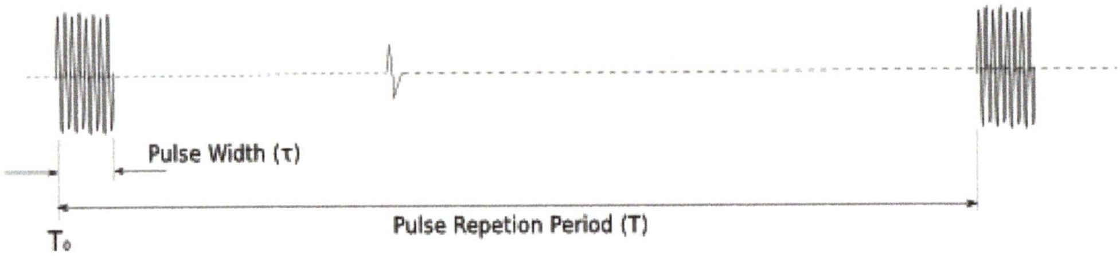

Carrier

The carrier is an RF signal, typically of microwave frequencies, which is usually (but not always) modulated to allow the system to capture the required data. In simple ranging radars, the carrier will be pulse modulated and in continuous wave systems, such as Doppler radar, modulation may not be required. Most systems use pulse modulation, with or without other supplementary modulating signals. Note that with pulse modulation, the carrier is simply switched on and off in sync with the pulses; the modulating waveform does not actually exist in the transmitted signal and the envelope of the pulse waveform is extracted from the demodulated carrier in the receiver. Although obvious when described, this point is often missed when pulse transmissions are first studied, leading to misunderstandings about the nature of the signal.

Pulse Width

The pulse width (τ) (or pulse duration) of the transmitted signal is to ensure that the radar emits sufficient energy to allow that the reflected pulse is detectable by its receiver. The amount of energy that can be delivered to a distant target is the product of two things; the output power of the transmitter, and the duration of the transmission. Therefore, pulse width constrains the maximum detection range of a target.

It also determines the range discrimination, that is the capacity of the radar to distinguish between two targets fairly close together. At *any* range, with similar azimuth and elevation angles and as viewed by a radar with an unmodulated pulse, the range discrimination is approximately equal in distance to half of the pulse duration.

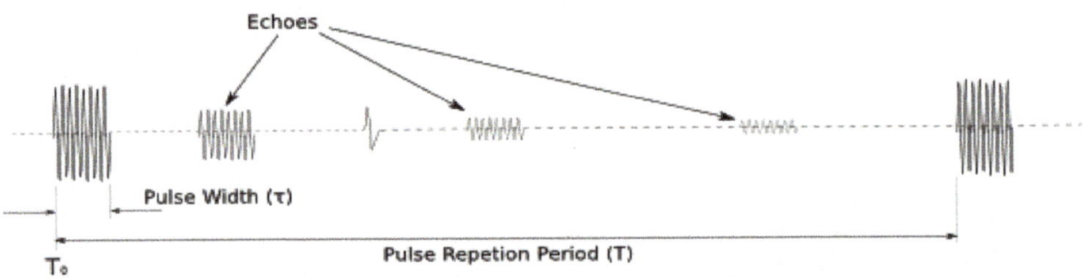

Pulse width also determines the dead zone at close ranges. While the radar transmitter is active, the receiver input is blanked to avoid the amplifiers being swamped (saturated) or, (more likely), damaged. A simple calculation reveals that a radar echo will take approximately 10.8 μs to return from a target 1 statute mile away (counting from the leading edge of the transmitter pulse (T_0), (sometimes known as transmitter main bang)). For convenience, these figures may also be expressed as 1 nautical mile in 12.4 μs or 1 kilometre in 6.7 μs. If the radar pulse width is 1 μs, then there can be no detection of targets closer than about 150 m, because the receiver is blanked.

All this means that the designer cannot simply increase the pulse width to get greater range without having an impact on other performance factors. As with everything else in a radar system, compromises have to be made to a radar system's design to provide the optimal performance for its role.

Pulse Repetition Frequency (PRF)

In order to build up a discernible echo, most radar systems emit pulses continuously and the repetition rate of these pulses is determined by the role of the system. An echo from a target will therefore be 'painted' on the display or integrated within the signal processor every time a new pulse is transmitted, reinforcing the return and making detection easier. The higher the PRF that is used, then the more the target is painted. However, with the higher PRF the range that the radar can "see" is reduced. Radar designers try to use the highest PRF possible commensurate with the other factors that constrain it, as described below.

There are two other facets related to PRF that the designer must weigh very carefully; the beamwidth characteristics of the antenna, and the required periodicity with which the radar must sweep the field of view. A radar with a 1° horizontal beamwidth that sweeps the entire 360° horizon every 2 seconds with a PRF of 1080 Hz will radiate 6 pulses over each 1-degree arc. If the receiver needs at least 12 reflected pulses of similar amplitudes to achieve an acceptable probability of detection, then there are three choices for the designer: double the PRF, halve the sweep speed, or double the beamwidth. In reality, all three choices are used, to varying extents; radar design is all about compromises between conflicting pressures.

Staggered PRF

Staggered PRF is a transmission process where the time between interrogations from radar changes slightly, *in a patterned and readily-discernible repeating manner*. The change of repetition frequency allows the radar, on a pulse-to-pulse basis, to differentiate between returns from its own transmissions and returns from other radar systems with the same PRF and a similar radio frequency. Consider a radar with a constant interval between pulses; target reflections appear at a relatively constant range related to the flight-time of the pulse. In today's very crowded radio spectrum, there may be many other pulses detected by the receiver, either directly from the transmitter or as reflections from elsewhere. Because their apparent "distance" is defined by measuring their time relative to the last pulse transmitted by "our" radar, these "jamming" pulses could appear at any apparent distance. When the PRF of the "jamming" radar is very similar to "our" radar, those apparent distances may be very slow-changing, just like real targets. By using stagger, a radar designer can force the "jamming" to jump around erratically in apparent range, inhibiting integration and reducing or even suppressing its impact on true target detection.

Without staggered PRF, any pulses originating from another radar on the same radio frequency might appear stable in time and could be mistaken for reflections from the radar's own transmission. With staggered PRF the radar's own targets appear stable in range in relation to the transmit pulse, whilst the 'jamming' echoes may move around in apparent range (uncorrelated), causing them to be rejected by the receiver. Staggered PRF is only one of several similar techniques used for this, including jittered PRF (where the pulse timing is varied in a less-predictable manner), pulse-frequency modulation, and several other similar techniques whose principal purpose is to reduce the probability of unintentional synchronicity. These techniques are in widespread use in marine safety and navigation radars, by far the most numerous radars on planet Earth today.

Clutter

Clutter refers to radio frequency (RF) echoes returned from targets which are uninteresting to the radar operators. Such targets include natural objects such as ground, sea, precipitation (such as rain, snow or hail), sand storms, animals (especially birds), atmospheric turbulence, and other atmospheric effects, such as ionosphere reflections, meteor trails, and three body scatter spike. Clutter may also be returned from man-made objects such as buildings and, intentionally, by radar countermeasures such as chaff.

Some clutter may also be caused by a long radar waveguide between the radar transceiver and the antenna. In a typical plan position indicator (PPI) radar with a rotating antenna, this will usually be seen as a "sun" or "sunburst" in the centre of the display as the receiver responds to echoes from dust particles and misguided RF in the waveguide. Adjusting the timing between when the transmitter sends a pulse and when the receiver stage is enabled will generally reduce the sunburst without affecting the accuracy of the range, since most sunburst is caused by a diffused transmit pulse reflected before it leaves the antenna. Clutter is considered a passive interference source, since it only appears in response to radar signals sent by the radar.

Clutter is detected and neutralized in several ways. Clutter tends to appear static between radar scans; on subsequent scan echoes, desirable targets will appear to move, and all stationary echoes can be eliminated. Sea clutter can be reduced by using horizontal polarization, while rain is reduced with circular polarization (note that meteorological radars wish for the opposite effect, and therefore use linear polarization to detect precipitation). Other methods attempt to increase the signal-to-clutter ratio.

Clutter moves with the wind or is stationary. Two common strategies to improve measure or performance in a clutter environment are:

- Moving target indication, which integrates successive pulses and
- Doppler processing, which uses filters to separate clutter from desirable signals.

The most effective clutter reduction technique is pulse-Doppler radar with Look-down/shoot-down capability. Doppler separates clutter from aircraft and spacecraft using a frequency spectrum, so individual signals can be separated from multiple reflectors located in the same volume using velocity differences. This requires a coherent transmitter. Another technique uses a moving target indication that subtracts the receive signal from two successive pulses using phase to reduce signals from slow moving objects. This can be adapted for systems that lack a coherent transmitter, such as time-domain pulse-amplitude radar.

Constant False Alarm Rate, a form of Automatic Gain Control (AGC), is a method that relies on clutter returns far outnumbering echoes from targets of interest. The receiver's gain is automatically adjusted to maintain a constant level of overall visible clutter. While this does not help detect targets masked by stronger surrounding clutter, it does help to distinguish strong target sources. In the past, radar AGC was electronically controlled and affected the gain of the entire radar receiver. As radars evolved, AGC became computer-software controlled and affected the gain with greater granularity in specific detection cells.

Clutter may also originate from multipath echoes from valid targets caused by ground reflection, atmospheric ducting or ionospheric reflection/refraction (e.g., Anomalous propagation). This clutter type is especially bothersome since it appears to move and behave like other normal (point) targets of interest. In a typical scenario, an aircraft echo is reflected from the ground below, appearing to the receiver as an identical target below the correct one. The radar may try to unify the targets, reporting the target at an incorrect height, or eliminating it on the basis of jitter or a physical impossibility. Terrain bounce jamming exploits this response by amplifying the radar signal and directing it downward. These problems can be overcome by incorporating a ground map of the radar's surroundings and eliminating all echoes which appear to originate below ground or above a certain height. Monopulse can be improved by altering the elevation algorithm used at low elevation. In newer air traffic control radar equipment, algorithms are used to identify the false targets by comparing the current pulse returns to those adjacent, as well as calculating return improbabilities.

Sensitivity Time Control (STC)

STC is used to avoid saturation of the receiver from close in ground clutter by adjusting the attenuation of the receiver as a function of distance. More attenuation is applied to returns close in and is reduced as the range increases.

Unambiguous Range

Single PRF

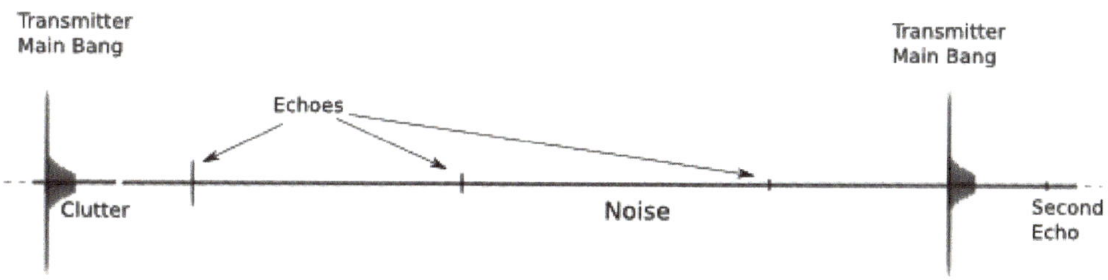

In simple systems, echoes from targets must be detected and processed before the next transmitter pulse is generated if range ambiguity is to be avoided. Range ambiguity occurs when the time taken for an echo to return from a target is greater than the pulse repetition period (T); if the interval between transmitted pulses is 1000 microseconds, and the return-time of a pulse from a distant target is 1200 microseconds, the apparent distance of the target is only 200 microseconds. In sum, these 'second echoes' appear on the display to be targets closer than they really are.

Consider the following example : if the radar antenna is located at around 15 m above sea level, then the distance to the horizon is pretty close, (perhaps 15 km). Ground targets further than this range cannot be detected, so the PRF can be quite high; a radar with a PRF of 7.5 kHz will return ambiguous echoes from targets at about 20 km, or over the horizon. If however, the PRF was doubled to 15 kHz, then the ambiguous range is reduced to 10 km and targets beyond this range would only appear on the display after the transmitter has emitted another pulse. A target at 12 km would appear to be 2 km away, although the strength of the echo might be much lower than that from a genuine target at 2 km.

The maximum non ambiguous range varies inversely with PRF and is given by:

$$\text{Range}_{\text{max unambiguous}} = \left(\frac{c}{2PRF}\right)$$

where *c* is the speed of light. If a longer unambiguous range is required with this simple system, then lower PRFs are required and it was quite common for early search radars to have PRFs as low as a few hundred Hz, giving an unambiguous range out to well in excess of 150 km. However, lower PRFs introduce other problems, including poorer target painting and velocity ambiguity in Pulse-Doppler systems.

Multiple PRF

Modern radars, especially air-to-air combat radars in military aircraft, may use PRFs in the tens-to-hundreds of kilohertz and stagger the interval between pulses to allow the correct range to be determined. With this form of staggered PRF, a *packet* of pulses is transmitted with a fixed interval between each pulse, and then another *packet* is transmitted with a slightly different interval. Target reflections appear at different ranges for each *packet*; these differences are accumulated and then simple arithmetical techniques may be applied to determine true range. Such radars may use repetitive patterns of *packets*, or more adaptable *packets* that respond to apparent target behaviors. Regardless, radars that employ the technique are universally coherent, with a very stable radio frequency, and the pulse *packets* may also be used to make measurements of the Doppler shift (a velocity-dependent modification of the apparent radio frequency), especially when the PRFs are in the hundreds-of-kilohertz range. Radars exploiting Doppler effects in this manner typically determine relative velocity first, from the Doppler effect, and then use other techniques to derive target distance.

Maximum Unambiguous Range

At its most simplistic, MUR (Maximum Unambiguous Range) for a Pulse Stagger sequence may be calculated using the TSP (Total Sequence Period). TSP is defined as the total time it takes for the Pulsed pattern to repeat. This can be found by the addition of all the elements in the stagger sequence. The formula is derived from the speed of light and the length of the sequence:

$$MUR = (c * 0.5 * TSP)$$

where c is the speed of light, usually in metres per microsecond, and TSP is the addition of all the positions of the stagger sequence, usually in microseconds. However, it should be noted that in a stagger sequence, some intervals may be repeated several times; when this occurs, it is more appropriate to consider TSP as the addition of all the unique intervals in the sequence.

Also, it is worth remembering that there may be vast differences between the MUR and the maximum range (the range beyond which reflections will probably be too weak to be detected), and that the maximum *instrumented* range may be *much* shorter than either of these. A civil marine radar, for instance, may have user-selectable maximum *instrumented* display ranges of 72, or 96 or rarely 120 nautical miles, in accordance with international law, but maximum unambiguous ranges of over 40,000 nautical miles and maximum detection ranges of perhaps 150 nautical miles. When such huge disparities are noted, it reveals that the primary purpose of staggered PRF is to reduce "jamming", rather than to increase unambiguous range capabilities.

The Radar Signal in the Frequency Domain

Pure CW radars appear as a single line on a Spectrum analyser display and when modulated with other sinusoidal signals, the spectrum differs little from that obtained with standard analogue modulation schemes used in communications systems, such as Frequency Modulation and consist of the carrier plus a relatively small number of sidebands. When the radar signal is modulated with a pulse train as shown above, the spectrum becomes much more complicated and far more difficult to visualise.

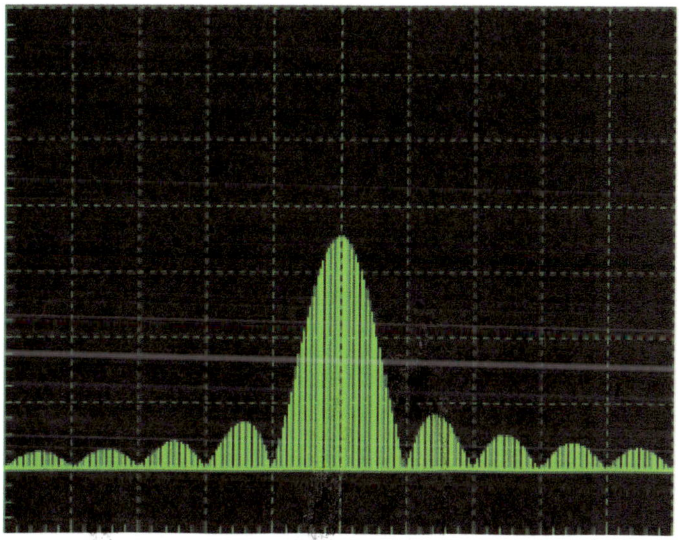

Basic Fourier analysis shows that any repetitive complex signal consists of a number of harmonically related sine waves. The radar pulse train is a form of square wave, the pure form of which consists of the fundamental plus all of the odd harmonics. The exact composition of the pulse train will depend on the pulse width and PRF, but mathematical analysis can be used to calculate all of the frequencies in the spectrum. When the pulse train is used to modulate a radar carrier, the typical spectrum shown on the left will be obtained.

Examination of this spectral response shows that it contains two basic structures. The Coarse Structure; (the peaks or 'lobes' in the diagram on the left) and the Fine Structure which contains the individual frequency components as shown below. The Envelope of the lobes in the Coarse Structure is given by: $\frac{1}{\pi f}$.

Note that the pulse width (τ) determines the lobe spacing. Smaller pulse widths result in wider lobes and therefore greater bandwidth.

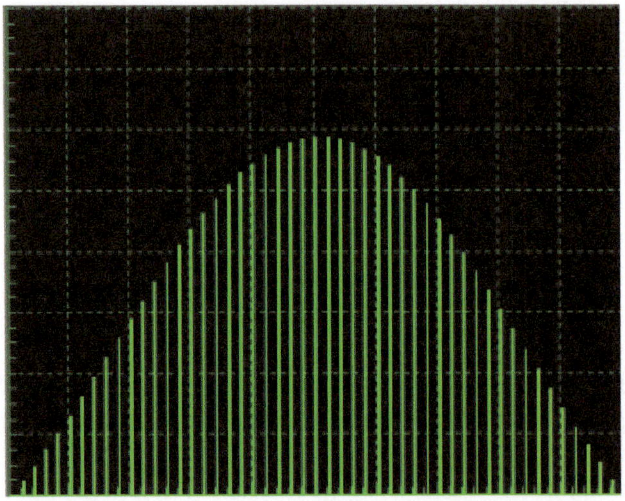

Examination of the spectral response in finer detail, as shown on the right, shows that the Fine Structure contains individual lines or spot frequencies. The formula for the fine structure is given by $\frac{N}{T}$ and since the period of the PRF (T) appears at the bottom of the fine spectrum equation, there will be fewer lines if higher PRFs are used. These facts affect the decisions made by radar designers when considering the trade-offs that need to be made when trying to overcome the ambiguities that affect radar signals.

Pulse Profiling

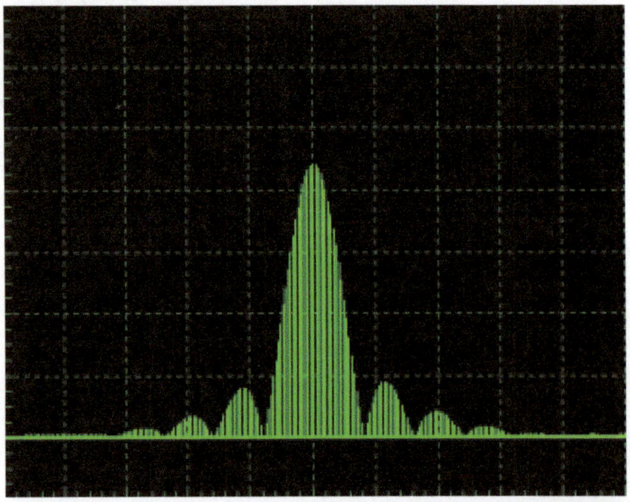

If the rise and fall times of the modulation pulses are zero, (e.g. the pulse edges are infinitely sharp), then the sidebands will be as shown in the spectral diagrams above. The bandwidth consumed by this transmission can be huge and the total power transmitted is distributed over many hundreds of spectral lines. This is a potential source of interference with any other device and frequency-dependent imperfections in the transmit chain mean that some of this power never arrives at the antenna. In reality of course, it is impossible to achieve such sharp edges, so in practical systems the sidebands contain far fewer lines than a perfect system. If the bandwidth can be limited to include relatively few sidebands, by rolling off the pulse edges intentionally, an efficient system can be realised with the minimum of potential for interference with nearby equipment. However, the trade-off of this is that slow edges make range resolution poor. Early radars limited the bandwidth through filtration in the transmit chain, e.g. the waveguide, scanner etc., but performance could be sporadic with unwanted signals breaking through at remote frequencies and the edges of the recovered pulse being indeterminate. Further examination of the basic Radar Spectrum shown above shows that the information in the various lobes of the Coarse Spectrum is identical to that contained in the main lobe, so limiting the transmit and receive bandwidth to that extent provides significant benefits in terms of efficiency and noise reduction.

Recent advances in signal processing techniques have made the use of pulse profiling or shaping more common. By shaping the pulse envelope before it is applied to the transmitting device, say to a cosine law or a trapezoid, the bandwidth can be limited at source, with less reliance on filtering. When this technique is combined with pulse compression, then a good compromise between efficiency, performance and range resolution can be realised. The diagram on the left shows the effect on the spectrum if a trapezoid pulse profile is adopted. It can be seen that the energy in the sidebands is significantly reduced compared to the main lobe and the amplitude of the main lobe is increased.

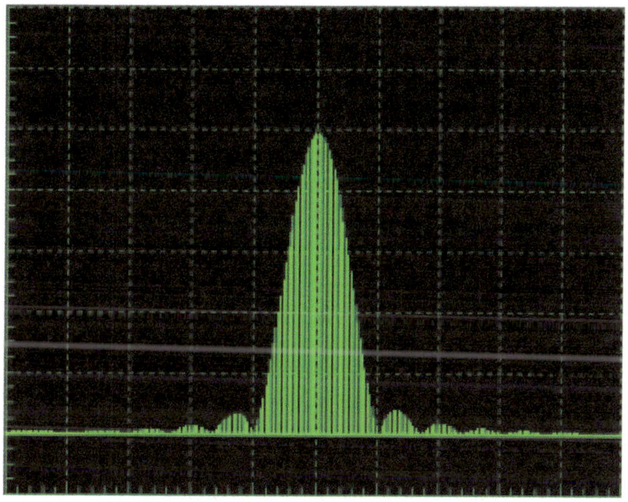

Similarly, the use of a cosine pulse profile has an even more marked effect, with the amplitude of the sidelobes practically becoming negligible. The main lobe is again increased in amplitude and the sidelobes correspondingly reduced, giving a significant improvement in performance.

There are many other profiles that can be adopted to optimise the performance of the system, but cosine and trapezoid profiles generally provide a good compromise between efficiency and resolution and so tend to be used most frequently.

Unambiguous Velocity

This is an issue only with a particular type of system; the Pulse-Doppler radar, which uses the Doppler effect to resolve velocity from the apparent change in frequency caused by targets that have net radial velocities compared to the radar device. Examination of the spectrum generated by a pulsed transmitter, shown above, reveals that each of the sidebands, (both coarse and fine), will be subject to the Doppler effect, another good reason to limit bandwidth and spectral complexity by pulse profiling.

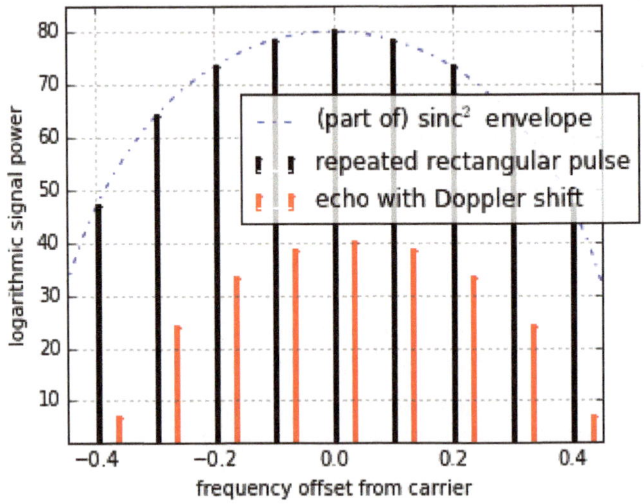

Consider the positive shift caused by the closing target in the diagram which has been highly simplified for clarity. It can be seen that as the relative velocity increases, a point will be reached where the spectral lines that constitute the echoes are hidden or aliased by the next sideband of the modulated carrier. Transmission of multiple pulse-packets with different PRF-values, e.g. staggered PRFs, will resolve this ambiguity, since each new PRF value will result in a new sideband position, revealing the velocity to the receiver. The maximum unambiguous target velocity is given by:

$$\pm\frac{cPRF}{4f}$$

Typical System Parameters

Taking all of the above characteristics into account means that certain constraints are placed on the radar designer. For example, a system with a 3 GHz carrier frequency and a pulse width of 1 µs will have a carrier period of approximately 333 ps. Each transmitted pulse will contain about 3000 carrier cycles and the velocity and range ambiguity values for such a system would be:

PRF	Velocity Ambiguity	Range Ambiguity
Low (2 kHz)	50 m/s	75 km
Medium (12 kHz)	300 m/s	12.5 km
High (200 kHz)	5000 m/s	750 m

Radar Display

A radar display is an electronic device to present radar data to the operator. The radar system transmits pulses or continuous waves of electromagnetic radiation, a small portion of which backscatter off targets (intended or otherwise) and return to the radar system. The receiver converts all received electromagnetic radiation into a continuous electronic analog signal of varying (or oscillating) voltage that can be converted then to a screen display.

Modern systems typically use some sort of raster scan display to produce a map-like image. Early in radar development, however, numerous circumstances made such displays difficult to produce. People ultimately developed several different display types.

Oscilloscopes

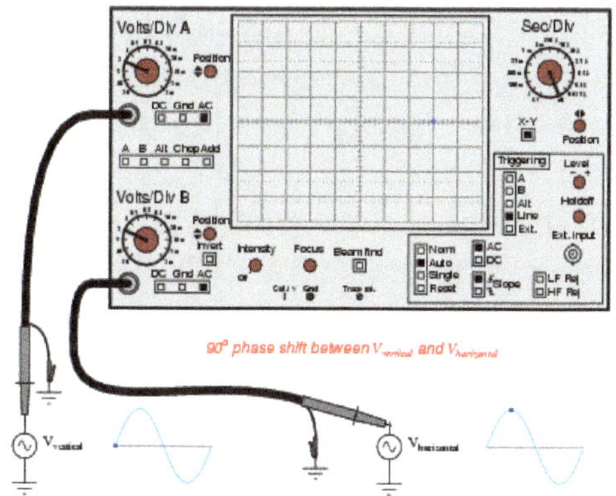

Oscilloscope attached to two sine-wave voltage sources, producing a circle pattern on the display.

Early radar displays used adapted oscilloscopes with various inputs. An oscilloscope generally receives three *channels* of varying (or oscillating) voltage as input and displays this information on a cathode ray tube. The oscilloscope amplifies the input voltages and sends them into two deflection magnets and to the electron gun producing a spot on the screen. One magnet displaces the spot horizontally, the other vertically, and the input to the gun increases or decreases the brightness of the spot. A bias voltage source for each of the three channels allows the operator to set a zero point.

In a radar display, the output signal from the radar receiver is fed into one of three input channels in the oscilloscope. Early displays generally sent this information to either X channel or Y channel to displace the spot on the screen to indicate a return. More modern radars typically used a rotating or otherwise moving antenna to cover a greater area of the sky, and in these cases, electronics, slaved to the mechanical motion of the antenna, typically moved the X and Y channels, with the radar signal being fed into the brightness channel.

A-Scope

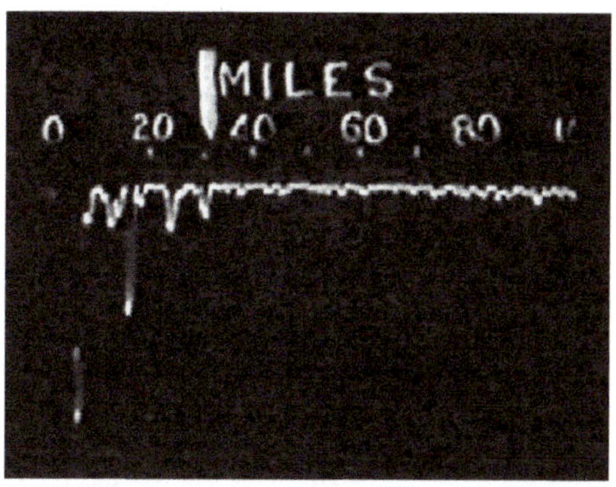

Chain Home is the canonical A-scope system. This image shows several target "blips" at ranges between 15 and 30 miles from the station. The large blip on the far left is the leftover signal from the radar's own transmitter, targets in this area could not be seen. The signal is inverted to make measurement simpler.

The original radar display, the A-scope or A-display, shows only the range, not the direction, to targets. Some people referred to these displays also as R-scope for *range scope*. A-scopes were used on the earliest radar systems during World War II, notably the seminal Chain Home (CH) system.

The primary input to the A-scope was the amplified return signal received from the radar, which was sent into the Y-axis of the display. Returns caused the spot to be deflected downward (or upward on some models), drawing vertical lines on the tube. These lines were known as a "blip" (or "pip"). The X-axis input was connected to a sawtooth voltage generator known as a time base generator that swept the spot across the display, timed to match the pulse repetition frequency of the radar. This spread out the vertical traces across the display according to the time they were received. Since the return time of the signal corresponds to twice the distance to the target divided by the speed of light, the distance along the axis directly indicates the range to any target. This was usually measured against a scale above the display.

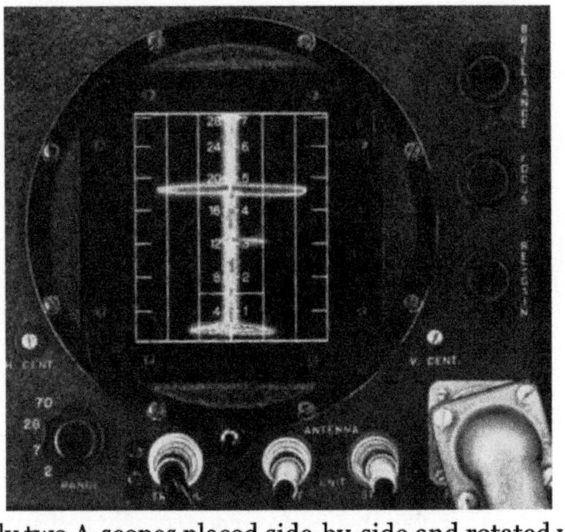

The L-scope was basically two A-scopes placed side-by-side and rotated vertically. By comparing the signal strength from two antennas, the rough direction of the blip could be determined. In this case there are two blips, a large one roughly centred, and a smaller one far to the right.

Early American and German radars used the J-scope, which resembled a circular version of the A-scope. These display range as an angle around the display face, as opposed to the linear distance along it. This arrangement allows greater accuracy in reading the range with the same sized display as an A-scope because the trace uses the full circumference rather than just the horizontal distance (so the time base is π times longer). An electro-mechanical version of the J-scope display remained common on consumer boating depth meters until recently.

Some early radars, notably versions of the Chain Home system, used the HR-scope, a modified A-scope. CH receivers were connected to a number of vertically displaced antennas, and by comparing the strength of a blip from each one, the operator could determine the vertical angle of the target, and then calculate its altitude. To aid this measurement, the systems were modified to allow the returns from two selected antennas to be displays at the same time, making it much easier to compare the blips. The name refers to "height-range".

A similarly modified version of the A-scope display was commonly used for air-to-air and ground-search radars, notably in AI radars and ASV radars - (Air-Surface Vessel). In this case, two receiver antennas were used in front of a common reflector, pointed slightly to the left and right of the aircraft centerline. Reception from both, using lobe switching, was sent to the left and right sides of a vertically oriented A-scope, and range could be measured as before. However, displacement of the target to the sides of the aircraft would result in the return being stronger on one side than the other, causing the blip on that side to be larger. This allowed the radar operator to easily indicate what direction to turn to intercept the target. These types of displays were sometimes referred to as ASV-scopes or L-scopes, although the naming was not universal.

Size of A-scope displays vary, but 5 to 7 inch diagonal was often used on a radar display. The 7JPx series of CRT's (7JP1, 7JP4 and 7JP7) was originally designed as an A-scope display CRT. Since the job of the A scope was to track an object, the phosphor used in the CRT has to have a long persistence (the trace remained on the display longer) from sweep to sweep. The P7 phosphor has the longest persistence so the 7JP7 was used.

B-Scope

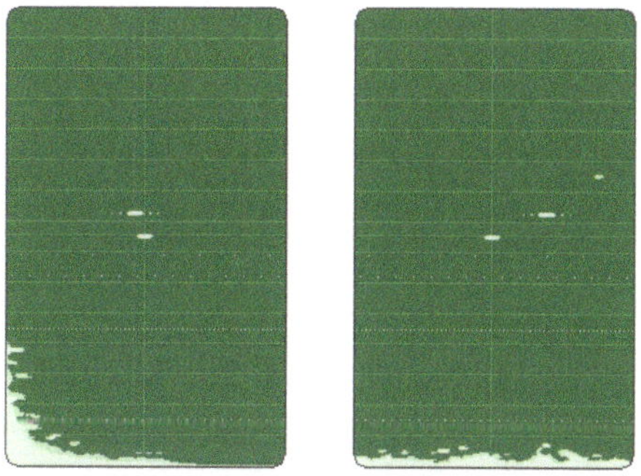

E-scope on the left and B-scope on the right

A B-scope or b-scan provides a 2-D "top down" representation of space, with the vertical axis typically representing range and the horizontal axis azimuth (angle). B-scope displays were common in airborne radars in the 1950s and 60s, which were mechanically scanned from side to side, and sometimes up and down as well. The B-scope's display represented a horizontal "slice" of the airspace on both sides of the aircraft out to the tracking angles of the radar. The spot was swept up the Y-axis in a fashion similar to the A-scope's X-axis, with distances "up" the display indicating greater range. This signal was mixed with a varying voltage being generated by a mechanical device that depended on the current horizontal angle of the antenna. The result was essentially an A-scope whose range line was rotated to point up, and then rotated back and forth about a zero point at the bottom of the display. The radio signal was sent into the intensity channel, producing a bright spot on the display indicating returns.

An E-scope is essentially a B-scope displaying range vs. elevation, rather than range vs. azimuth. They are identical in operation to the B-scope, the name simply indicating "elevation". E-scopes are typically used with height finding radars, which are similar to airborne radars but turned to scan vertically instead of horizontally, they are also sometimes referred to as "nodding radars" due to their antenna's motion. The display tube was generally rotated 90 degrees to put the elevation axis vertical in order to provide a more obvious correlation between the display and the "real world". These displays are also referred to as a Range-Height Indicator, or RHI, but were also commonly referred to (confusingly) as a B-scope as well.

The H-scope is another modification of the B-scope concept, but displays elevation as well as azimuth and range. The elevation information is displayed by drawing a second "blip" offset from the target indicator by a short distance, the *slope* of the line between the two blips indicates the elevation relative to the radar. For instance, if the blip were displaced directly to the right this would indicate that the target is at the same elevation as the radar. The offset is created by dividing the radio signal into two, then slightly delaying one of the signals so it appears offset on the display. The angle was adjusted by delaying the *time* of the signal via a delay, the length of the delay being controlled by a voltage varying with the vertical position of the antenna. This sort of elevation display could be added to almost any of the other displays, and was often referred to as a "double dot" display.

C-Scope

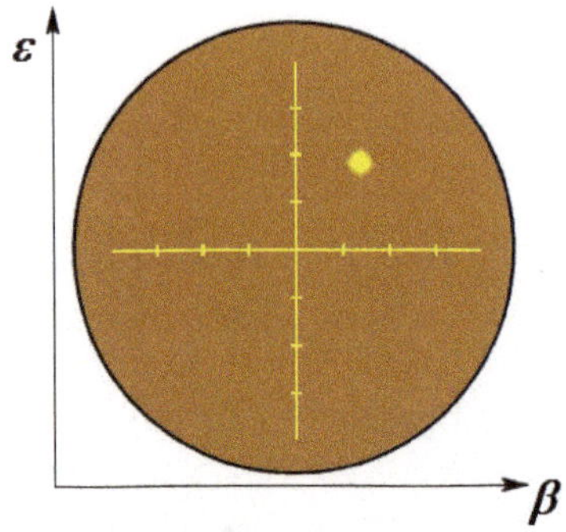

C-scope display. The target is above and to the right of the radar, but the range is not displayed.

A C-scope displays a "bullseye" view of azimuth vs. elevation. The "blip" was displayed indicating the direction of the target off the centreline axis of the radar, or more commonly, the aircraft or gun it was attached to. They were also known as "moving spot indicators", the moving spot being the target blip. Range is typically displayed separately in these cases, often as a number at the side of the display.

Almost identical to the C-scope is the G-scope, which overlays a graphical representation of the range to the target. This is typically represented by a horizontal line that "grows" out from the target indicator "blip" to form a wing-like diagram. The wings grew in length at shorter distances to indicate the target was closer. A "shoot now" range indicator is often supplied as well, typically consisting of two short vertical lines centered on either side of the middle of the display. To make an interception, the pilot guides his aircraft until the blip is centered, then approaches until the "wings" fill the area between the range markers. This display recreated a system commonly used on gunsights, where the pilot would dial in a target's wingspan and then fire when the wings filled the area inside a circle in their sight. This system allowed the pilot to estimate the range to the target. In this case, however, the range is being measured directly by the radar, and the display was mimicking the optical system to retain commonality between the two systems.

Plan Position Indicator

The PPI display provides a 2-D "all round" display of the airspace around a radar site. The distance out from the center of the display indicates range, and the angle around the display is the azimuth to the target. The current position of the radar antenna is typically indicated by a line extending from the center to the outside of the display, which rotates along with the antenna in realtime. It is essentially a B-scope extended to 360 degrees. The PPI display is typically what people think of as a radar display in general, and was widely used in air traffic control until the introduction of raster displays in the 1990s.

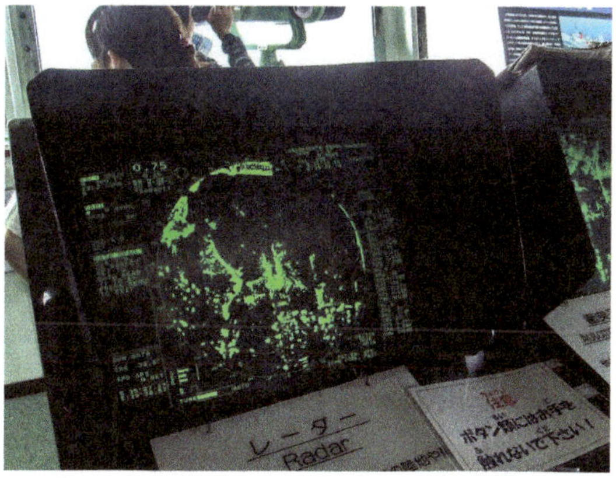

This image shows a modern PPI display in use, with the islands and ground surrounding the ship in green. The current direction of the radar can be seen as the dotted line pointing northwest.

PPI displays are actually quite similar to A-scopes in operation, and appeared fairly quickly after the introduction of radar. As with most 2D radar displays, the output of the radio receiver was attached to the intensity channel to produce a bright dot indicating returns. In the A-scope a saw-tooth voltage generator attached to the X-axis moves the spot across the screen, whereas in the PPI the output of two such generators is used to rotate the line around the screen. Some early systems were mechanical, physically spinning the deflection magnets, but the electronics needed to do this in a "solid-state" fashion were not particularly complex, and were in use in the early 1940s.

Beta Scan Scope

A Beta Scan display.

The specialist Beta Scan Scope was used for precision approach radar systems. It displays two lines on the same display, the upper one (typically) displaying the vertical approach (the glideslope), and the lower one the horizontal approach. A marker indicates the desired touchdown point on the runway, and often the lines are angled towards the middle of the screen to indicate this location. A single aircraft's "blip" is also displayed, superimposed over both lines, the signals being generated from separate antennas. Deviation from the centerline of the approach can be seen and easily relayed to the pilot.

In the image, the long lines running diagonally across the screen represents the middle of the glideslope, the vertical bars are distance markers, and the dashed lines (just visible) represent the "maximum safe distance" an aircraft can stray from the glideslope. The distance between these markers and the centerline decreases closer to the touchdown point.

Plan Position Indicator

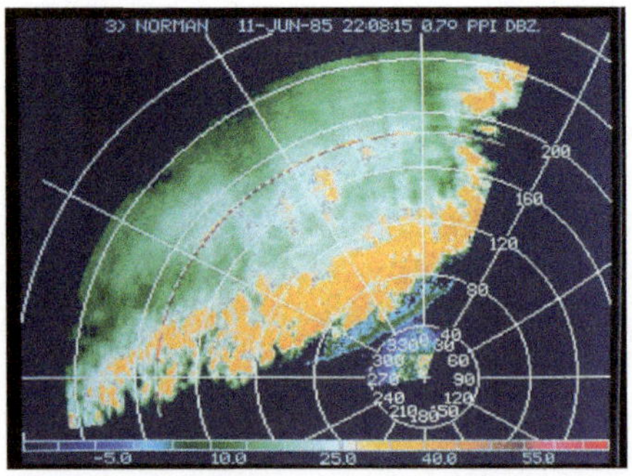

Image of a thunderstorm line (in dBZ) seen on a 0.7-degree elevation PPI (NOAA)

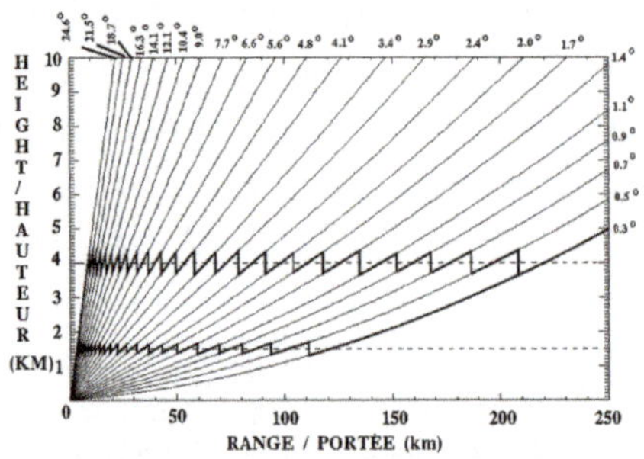

Diagram showing the evolution of the height above ground, in kilometers, with the distance to the radar for the 24 PPI angles used on the Canadian weather radars (curved lines)

The plan position indicator (PPI), is the most common type of radar display. The radar antenna is usually represented in the center of the display, so the distance from it and height above ground can be drawn as concentric circles. As the radar antenna rotates, a radial trace on the PPI sweeps in unison with it about the center point.

Description

The radar antenna sends pulses while rotating 360 degrees around the radar site at a fixed elevation angle. It can then change angle or repeat at the same angle according to the need. Return echoes from targets are received by the antenna and processed by the receiver and the most direct display of those data is the PPI.

It is to be noted that the height of the echoes increases with the distance to the radar, as represented in the adjacent image. This change is not a straight line but a curve as the surface of the Earth is curved and *sinks* below the radar horizon. For fixed-site installations, north is usually represented at the top of the image. For moving installations, such as small ship and aircraft radars, the top may represent the bow or nose of the ship or aircraft, *i.e.*, its heading (direction of travel) and this is usually represented by a lubber line. Some systems may incorporate the input from a gyrocompass to rotate the display and once again display north as "up".

Also, the signal represented is the reflectivity at only one elevation of the antenna, so it is possible to have many PPIs at one time, one for each antenna elevation.

History

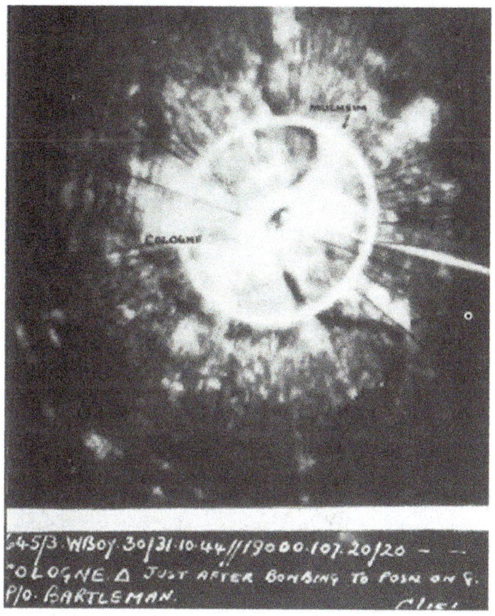

A photograph of an H2S PPI display taken during an attack on Cologne. The annotations were added later for post-attack analysis. The Rhine River can clearly be seen.

The PPI display was first used prior to the start of the Second World War in a Jagdschloss experimental radar system outside Berlin. The first production PPI was devised at the Telecommunications Research Establishment, UK and was first introduced in the H2S radar blind-bombing system of World War II.

Originally, data was displayed in real time on a cathode ray tube, and thus the only way to store the information received was by taking a photograph of the screen.

Philo Taylor Farnsworth, the American inventor of all-electronic television in September 1927, contributed to this in an important way. Farnsworth refined a version of his picture tube (cathode ray tube, or CRT) and called it an "Iatron;" generically known as a storage tube. It could store an image for milliseconds to minutes and even hours. One version that kept an image alive about a second before fading proved to be useful for radar. This slow-to-fade display tube was used by air traffic controllers from the very beginning of radar usage.

With the development of more sophisticated radar systems, it became possible to digitize data and store it in memory, allowing access at a later date.

Uses

The PPI is used in many domains involving display of range and positioning, especially in radars, including air traffic control, ship navigation, meteorology, on board ships and aircraft etc. PPI displays are also used to display sonar data, especially in underwater warfare. However, because the speed of sound in water is very slow compared to microwaves in air, a sonar PPI has an expanding circle that starts with each transmitted "ping" of sound. In meteorology, a competing display system is the CAPPI (Constant Altitude Plan Position Indicator) when a multi-angle scan is available.

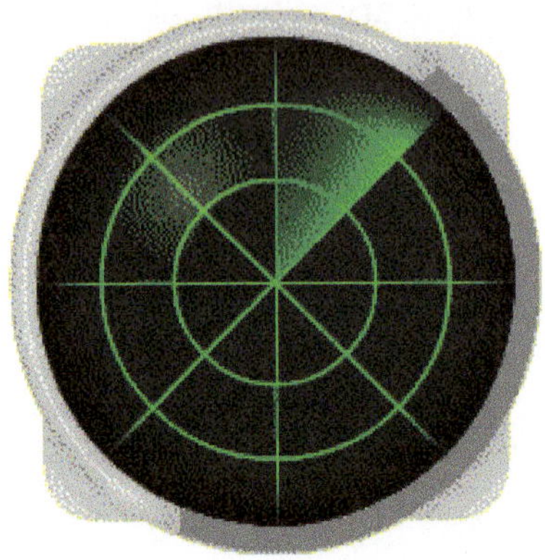

A Plan Position Indicator radar display

Using computers to process data, modern sonar and lidar installations can mimic radar PPI displays too.

Constant Altitude Plan Position Indicator

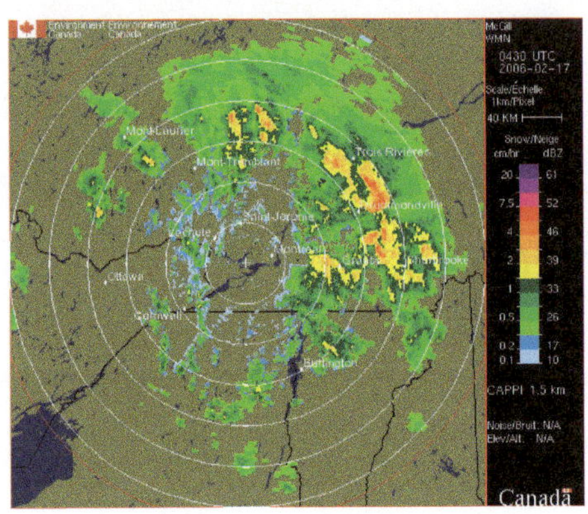

Reflectivity Cappi at 1.5 km altitude from Montréal radar (Source Environment Canada)

The constant altitude plan position indicator, better known as CAPPI, is a radar display which gives a horizontal cross-section of data at constant altitude. It has been developed by McGill University in Montreal by the *Stormy Weather Group* to circumvent some problems with the PPI:

- Altitude changing with distance to the radar.
- Ground echoes problems near the radar.

Definition and history

In 1954, McGill University got a new radar(CPS-9) which had a better resolution and used FASE (Fast Azimuth Slow Elevation) to program multi-angle soundings of the atmosphere.

In 1957, Langleben and Gaherty developed a scheme with FASE to keep only the data at a certain height at each angle and scan on 360 degrees. If we look at the diagram, each angle of elevation or PPI has data at height X at a certain distance from the radar. Using the data at the right distance, one forms an annular ring of data at height X. Assembling all the rings coming from the different angles gives you the CAPPI.

The CAPPI is composed of data from each angle that is at the height requested for the cross-section (bold lines in zig-zag on the left diagram). In the early days, the scan data collected where shown directly on the cathodic screen and a photo sensitive device captured each ring as it was completed. Then all those photographed rings were assembled. By 1958, East developed a real time assembly instead of a delayed one. By the mid 1970s, computer developments made it possible to gather data in electronic form and make CAPPIs more easily.

Today, weather radars collect in real-time data on a large number of angles. Many countries such as Canada, UK and Australia, scan a large enough number of angles with their radars to have an almost continuous vertical view (taking into account the radar beam width) and produce CAPPIs. Other countries, like France and United States, use fewer angles and prefer PPIs or composite of maximum reflectivities above a point.

Usage

Above right is an example of a CAPPI at 1.5 km altitude. Looking at the diagram of angles, that depending on the height of the CAPPI, there comes a distance where no data is available. The portion beyond this distance on a CAPPI is then showing data from the lowest PPI. The higher is the CAPPI above ground, the smaller is that PPI zone.

References

- Riquelme, A.J.; Abellán, A.; Tomás, R. (2015). "Discontinuity spacing analysis in rock masses using 3D point clouds". Engineering Geology. 195: 185–195. doi:10.1016/j.enggeo.2015.06.009
- Judkins, Phil. "Making Vision into Power", International Journal of Engineering and Technology, Vol 82, No 1 (January 2012), pp. 103-104
- Cracknell, Arthur P.; Hayes, Ladson (2007) [1991]. Introduction to Remote Sensing (2 ed.). London: Taylor and Francis. ISBN 0-8493-9255-1. OCLC 70765252
- Susan Cobb (October 29, 2004). "Weather radar development highlight of the National Severe Storms Laboratory first 40 years". NOAA Magazine. NOAA. Retrieved 2009-03-07
- van Keuren, D.K. (1997). "Science Goes to War: The Radiation Laboratory, Radar, and Their Technological Consequences". Reviews in American History. 25: 643–647. doi:10.1353/rah.1997.0150
- "Rediscovering the lost archaeological landscape of southern New England using airborne light detection and ranging (LiDAR)". Journal of Archaeological Science. 43: 9–20. doi:10.1016/j.jas.2013.12.004. Retrieved 2016-02-22
- Vosselman, George; Maas, Hans-Gerd (2012). Airborne and terrestrial laser scanning. Whittles Publishing. ISBN 978-1-904445-87-6
- "Information about Canadian radar network". The National Radar Program. Environment Canada. 2002. Archived from the original on 2004-06-29. Retrieved 2006-06-14
- C. Wilson, Jerry. "Using Airborne hydrographic lidar to support mapping of California's waters". OCEANS 2008 – MTS/IEEE Kobe Techno-Ocean, ISBN 978-1-4244-2126-8
- "LIDAR—Light Detection and Ranging—is a remote sensing method used to examine the surface of the Earth". NOAA. Archived from the original on June 4, 2013. Retrieved June 4, 2013
- Tang, Lina; Shao, Guofan (2015-06-21). "Drone remote sensing for forestry research and practices". Journal of Forestry Research. 26 (4): 791–797. ISSN 1007-662X. doi:10.1007/s11676-015-0088-y

- Mei, Liang. "Remote sensing of atmospheric aerosol and gas using scheimpflug lidar based on diode lasers". Progress in Electromagnetic Research Symposium (PIERS). ISBN 978-1-5090-6094-8
- "Filipino turns ordinary car into autonomous vehicle – Motioncars | Motioncars". Motioncars.inquirer.net. 2015-05-25. Retrieved 2016-02-22
- Gigli, G.; Casagli, N. (2011). "Semi-automatic extraction of rock mass structural data from high resolution LIDAR point clouds". International Journal of Rock Mechanics and Mining Sciences. 48: 187–198. doi:10.1016/j.ijrmms.2010.11.009
- Medina A, Gayá F, Pozo F (2006). "Compact laser radar and three-dimensional camera". J. Opt. Soc. Am. A. 23: 800–805. Bibcode:2006JOSAA..23..800M. doi:10.1364/josaa.23.000800

2

Understanding Radar Systems

To develop a better understanding of radar systems, it is important to understand monopulse radar, passive radar, pulse-Doppler radar, space-based radar and radar tracker. The pulse-Doppler radar uses the Doppler effect and provides detection of objects below a flying aircraft. This chapter is an overview of the subject matter incorporating all the major aspects of weather radars.

Bistatic Radar

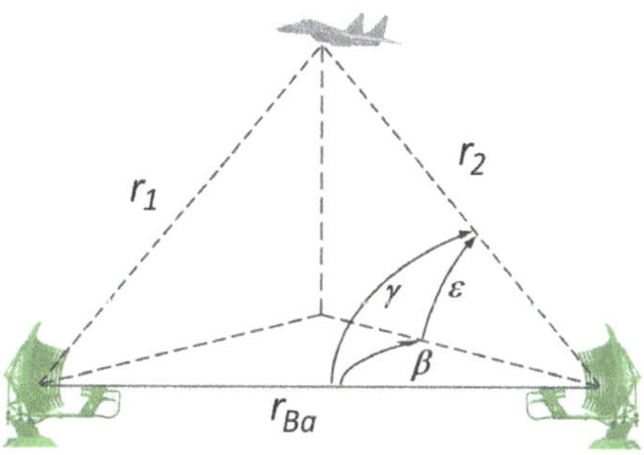

Bistatic Radar block diagram.

Bistatic radar is the name given to a radar system which comprises a transmitter and receiver which are separated by a distance that is comparable to the expected target distance. Conversely, a radar in which the transmitter and receiver are collocated is called a monostatic radar. A system containing multiple spatially diverse monostatic radar or bistatic radar components with a shared area of coverage is called multistatic radar. Many long-range air-to-air and surface-to-air missile systems use semi-active radar homing which is a form of bistatic radar.

Specific Classes of Bistatic Radar

Pseudo-monostatic Radars

Some radar systems may have separate transmit and receive antennas, but if the angle subtended between transmitter, target and receiver (the bistatic angle) is close to zero, then they would still be regarded as monostatic or pseudo-monostatic. For example, some very long range HF radar systems may have a transmitter and receiver which are separated by a few tens of kilometres for electrical isolation, but as the expected target range is of the order 1000-3500 km, they are not considered to be truly bistatic and are referred to as pseudo-monostatic.

Forward Scatter Radars

In some configurations, bistatic radars may be designed to operate in a fence-like configuration, detecting targets which pass between the transmitter and receiver, with the bistatic angle near 180 degrees. This is a special case of bistatic radar, known as a forward scatter radar, after the mechanism by which the transmitted energy is scattered by the target. In forward scatter, the scattering can be modeled using Babinet's principle and is a potential countermeasure to stealth aircraft as the radar cross section (RCS) is determined solely by the silhouette of the aircraft seen by the transmitter, and is unaffected by stealth coatings or shapings. The RCS in this mode is calculated as $\sigma=4\pi A^2/\lambda^2$, where A is the silhouette area and λ is the radar wavelength. However, target may vary from place to place location and tracking is very challenging in forward scatter radars, as the information content in measurements of range, bearing and Doppler becomes very low (all these parameters tend to zero, regardless of the location of the target in the fence).

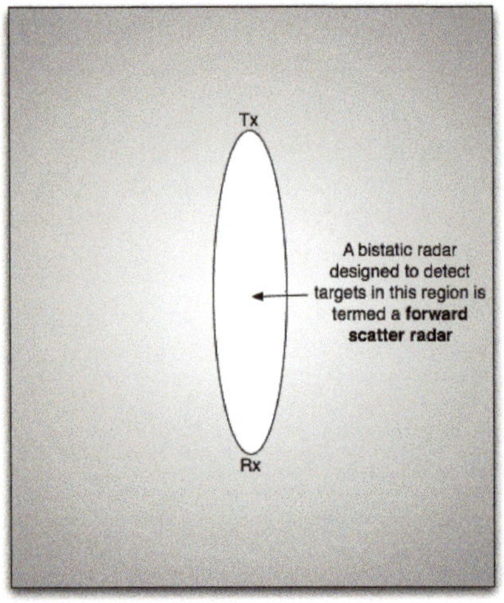

Illustration of forward scatter geometry

Multistatic Radar

A multistatic radar system is one in which there are at least three components - for example, one receiver and two transmitters, or two receivers and one transmitter, or multiple receivers and multiple transmitters. It is a generalisation of the bistatic radar system, with one or more receivers processing returns from one or more geographically separated transmitter.

Passive Radar

A bistatic or multistatic radar that exploits non-radar transmitters of opportunity is termed a passive coherent location system or passive covert radar.

Any radar which does not send active electronic magnetic pulse is known as passive radar. Passive coherent location also known as PCL is a special type of passive radar, which exploits the transmitters of opportunity especially the commercial signals in the environment.

Advantages and Disadvantages

The principal advantages of bistatic and multistatic radar include:

- Lower procurement and maintenance costs (if using a third party's transmitter)

- Operation without a frequency clearance (if using a third party's transmitter)
- Covert operation of the receiver
- Increased resilience to electronic countermeasures as waveform being used and receiver location are potentially unknown
- Possible enhanced radar cross section of the target due to geometrical effects
- Separate receiver is very light and mobile while transmitter can be very heavy and powerful (surface-to-air missile).

The principal disadvantages of bistatic and multistatic radar include:

- System complexity
- Costs of providing communication between sites
- Lack of control over transmitter (if exploiting a third party transmitter)
- Harder to deploy
- Reduced low-level coverage due to the need for line-of-sight from several locations.

Monopulse Radar

Monopulse radar is a radar system that compares the received signal from a single radar pulse against itself in order to compare the signal as seen in multiple directions, polarizations, or other differences. The most common form is an adaptation of conical scanning radar which compares the return from two directions to directly measure the location of the target. This avoids problems in decoding conventional conical scanning systems, which can be confused by rapid changes in signal strength. The system also makes jamming more difficult. Most radars designed since the 1960s are monopulse systems.

Description

Conical Scan

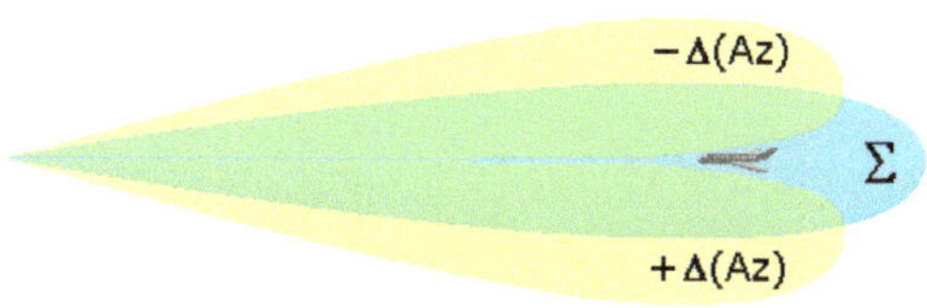

The basic idea behind conical scanning and monopulse radars is to use a beam that is slightly spread out across the antenna's centreline.

Conical scanning is not considered to be a form of monopulse radar, but the following summary provides background that can aid understanding.

Conical scan systems send out a signal slightly to one side of the antenna's boresight, and then rotating the feed horn to make the lobe rotate around the boresight line. A target centered on the boresight is always slightly illuminated by the lobe, and provides a strong return. If the target is to one side, it will be illuminated only when the lobe is pointed in that general direction, resulting in a weaker signal overall (or a flashing one if the rotation is slow enough). This varying signal will reach a maximum when the antenna is rotated so it is aligned in the direction of the target, by looking for this maximum and moving the antenna in that direction, a target can be automatically tracked.

One problem with this approach is that radar signals often change in amplitude for reasons that have nothing to do with beam position. Over the period of a few tenths of seconds, for instance, changes in target heading, rain clouds and other issues can dramatically affect the returned signal. Since conical scanning systems depend on the signal growing or weakening due only to the position of the target relative to the beam, such changes in reflected signal can cause it to be "confused" about the position of the target within the beam's scanning area.

Jamming a conical scanner is also relatively easy. The jammer simply has to send out signals on the radar's frequency with enough strength to make it think that was the strongest return. In this case a series of random short bursts of signal will appear to be a series of targets in different locations within the beam. Jamming of this sort can be made more effective by timing the signals to be the same as the rotational speed of the feed, but broadcast at a slight delay, which results in a second strong peak within the beam, with nothing to distinguish the two. Jammers of this sort were deployed quite early, the British used them during World War II against the German conical-scanning Würzburg radar.

Monopulse Basics

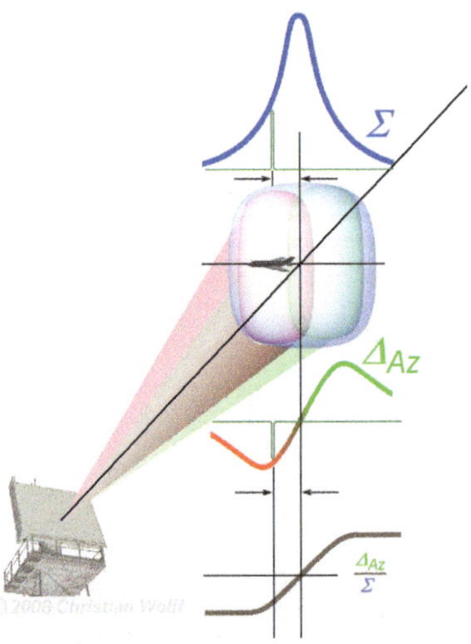

Monopulse beam of a phased array antenna split in two lobes.

Monopulse radars are similar in general construction to conical scanning systems, but split the beam into parts and then send the two resulting signals out of the antenna in slightly different directions. When the reflected signals are received they are amplified separately and compared to each other, indicating which direction has a stronger return, and thus the general direction of the target relative to the boresight. Since this comparison is carried out during one pulse, which is typically a few microseconds, changes in target position or heading will have no effect on the comparison.

Making such a comparison requires that different parts of the beam be distinguished from each other. Normally this is achieved by splitting the pulse into two parts and polarizing each one separately before sending it to a set of slightly off-axis feed horns. This results in a set of lobes, usually two, overlapping on the boresight. These lobes are then rotated as in a normal conical scanner. On reception the signals are separated again, and then one signal is inverted in power and the two are then summed (Σ in the image). If the target is to one side of the boresight the resulting sum will be positive, if it's on the other, negative.

If the lobes are closely spaced, this signal can produce a high degree of pointing accuracy within the beam, adding to the natural accuracy of the conical scanning system. Whereas classical conical scan systems generate pointing accuracy on the order of 0.1 degree, monopulse radars generally improve this by a factor of 10, and advanced tracking radars like the AN/FPS-16 are accurate to 0.006 degrees. This is an accuracy of about 10 m at a distance of 100 km.

Jamming resistance is greatly improved over conical scanning. Filters can be inserted to remove any signal that is either unpolarized, or polarized only in one direction. In order to confuse such a system, the jamming signal would have to duplicate the polarization of the signal as well as the timing, but since the aircraft receives only one lobe, determining the precise polarization of the signal is difficult. Against monopulse systems, ECM has generally resorted to broadcasting white noise to simply blind the radar, instead of attempting to produce false localized returns.

Antenna Sampling

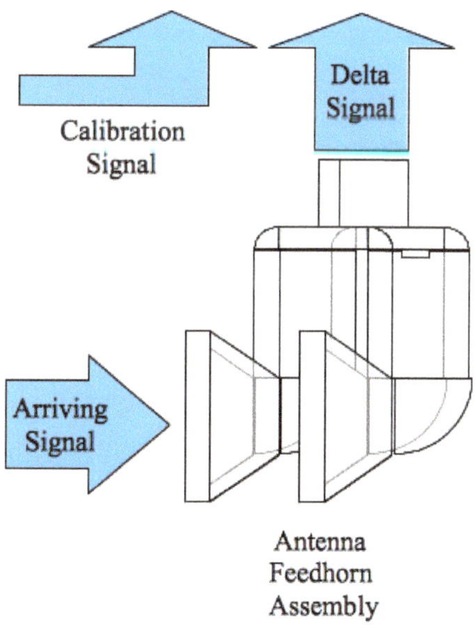

Monopulse antennas produce a sum signal and two delta signals. This allows angular measurement to be performed using a single receive pulse. The sum signal usually passes back down the waveguide used to send the transmit pulse. The two delta signals are elevation (up-down) and traverse (left-right).

The sum signal corresponds with the antenna beam along center-line of the antenna. The delta signals are pairs of beams that are adjacent to the center-line of the sum antenna beam. The delta beam measurements produce plus or minus values depending upon the quadrant.

Quadrants	LEFT	RIGHT
UP	QUADRANT II: +ΔEl -ΔAz	QUADRANT I: +ΔEl +ΔAz
DOWN	QUADRANT III: -ΔEl -ΔAz	QUADRANT IV: -ΔEl +ΔAz

The sum signal is created by a feedhorn structure positioned to maximize signal at the center of the antenna beam. The delta RF signals are created by pairs of antenna feed-horns located adjacent to the sum feed-horn (sum feed-horn not shown in the figure). The output of each pair of delta feed-horns are added together, and this creates zero output signal when the incoming RF signal is located at the center of the antenna beam. The signal strength from each delta beam increases as the aircraft drifts further away from the antenna center-line.

For the waveguide image that is shown, a horizontally polarized RF signal arrives at the two feed horns to produce a left/right delta signal. The arriving energy from the RF wavefront is launched into both waveguide feedhorns. The RF signal from both feedhorns travels up the waveguide where the signals from the left and right feedhorn are combined. The combiner performs a mathematical subtraction on the electrical signals arriving from the feedhorns. That subtraction produces the delta signal. A similar feedhorn configuration is used to produce the up/down delta signal. The waveguide assembly can be used by itself. For a high gain antenna, the feedhorn assembly is located facing the reflecting surface at or near the focal point.

For the waveguide image that is shown, the sum signal would be created by a single waveguide feedhorn centered between the two feedhorns that are shown.

The sum and delta radio frequency signals are converted to a lower frequency in the receiver where sampling takes place. A signal processor produces the error signal using these samples.

The + or - value for each delta signal is created by phase shift of 0 degrees or 180 degrees when compared with the sum signal. A calibration signal is injected into the receive path when the radar is idle, and this establishes a known phase shift between different microwave signal paths (quiescent mode).

The angle error is created from the delta signal by performing a complex ratio. This is done for the left/right delta beams, and this is also done for the up/down delta beams (two ratios). An explanation of how real and imaginary parts are used with RADAR can be found in the description of Pulse Doppler.

$$\text{Error} = \frac{\dfrac{(\text{Real Delta}) + i(\text{Imaginary Delta})}{(\text{Real Sum}) + i(\text{Imaginary Sum})}}{\dfrac{(\text{Real Delta Cal}) + i(\text{Imaginary Delta Cal})}{(\text{Real Sum Cal}) + i(\text{Imaginary Sum Cal})}}$$

The result is a signed number. The outcome of the calibration process is to rotate the complex antenna angle error vector onto the real axis to reduce signal processing losses.

The angle error is used to make an adjustment to position the target along the centerline of the antenna. On mechanically steered radar, the vertical angle error drives a motor that moves the antenna up or down, and the horizontal angle error drives a motor that steers the antenna left or right. On a missile, the angle error is an input to the guidance system that positions the guidance fins that rotate the body of the missile so that the target is in the centerline of the antenna.

A wheel, mirror and a light can be used to visualize real and imaginary described in this equation. The mirror is placed at a 45 degree angle above the wheel so that you can see the front and top of the wheel at the same time. The light is attached to the wheel so that you can see the wheel when the room lights are turned off. You sit directly in front of the wheel while a friend rotates the wheel. The view of the front of the wheel (real) and the top of the wheel (imaginary) tell you the position of the wheel.

Pairs of real and imaginary values form a complex number explained as real and imaginary parts.

Dynamic calibration is needed when there are long waveguide runs between the antenna and first down converter. This compensates for temperature changes that alter the size and length of wave-guide, which will cause phase variations that produce incorrect angle error signals for long wave-guide runs. The Cal term is created by injecting a calibration signal into the receive waveguide while the system is not active (sum and delta). The angle error of the calibration signal is used to evaluate angle error during normal operation. Antenna tuning is used to make adjustments that create the desired error signal when the antenna is calibrated on an antenna range.

When the waveguide run is short between the antenna and receiver, the calibration signal can be omitted and the calibration term can be set to a fixed value. A fixed value may also be stored for systems with long wave-guide runs to allow degraded operation when RF calibration cannot be performed. The waveguide assembly may need to be tuned using an antenna range to obtain consistent results.

Antenna Positioning

Tracking systems produce constant aircraft position information, and the antenna position is part of this information. Antenna error signals are used to create feedback as part of a RADAR system that can track aircraft.

The horizontal signal and the vertical signal created from antenna RF samples are called angle errors. These angle error signals indicate the angular distance between the center of the antenna beam and the position of the aircraft within the antenna beam.

For a mechanically steered antenna, the horizontal signal and vertical signal are used to create a drive signal that creates torque for two antenna positioning motors. One motor moves the antenna left/right. The other motor drives the antenna up/down. The result is to move the antenna position so that the center of the antenna beam remains aimed directly at the aircraft even when the aircraft is moving perpendicular to the antenna beam.

For a track while scan radar, position and velocity is maintained for multiple aircraft. The last position of the aircraft is coasted using the velocity, and that information is used to direct a beam of energy toward the aircraft. The monopulse angle error information that is received is used to adjust the position and velocity data for the aircraft. This is a common mode with phased array radar systems.

Amplitude-Comparison Monopulse provides an explanation of the antenna signals involved in this process.

Doppler

Doppler effect can be used to separate different objects based on speed. Pulse Doppler RADAR signal processing uses this technique. This is combined with conical scanning or monopulse to improve track reliability. It is necessary to separate the object signal from the interference to avoid being pulled off the object. This avoids problems where the system is fooled by aircraft flying too close to the surface of the earth or aircraft flying through clouds.

Conical scan and monopulse antennas are susceptible to interference from weather phenomenon and stationary objects. The resulting interference can produce feedback signals that move the antenna beam away from the aircraft. This can produce an unreliable antenna position when the antenna is aimed too near the ground or too near to heavy weather. Systems with no Pulse Doppler tracking mode may remain aimed at irrelevant objects like trees or clouds. Constant operator attention is required when there is no Doppler signal processing.

History

Monopulse radar was extremely "high tech" when it was first introduced by Robert M. Page in 1943 in a Naval Research Laboratory experiment. As a result, it was very expensive, labor-intensive due to complexity, and less reliable. It was only used when extreme accuracy was needed that justified the cost. Early uses included the Nike Ajax missile, which demanded very high accuracy, or for tracking radars used for measuring various rocket launches. An early monopulse radar development, in 1958, was the AN/FPS-16, on which NRL and RCA collaborated. The earliest version, XN-1, utilised a metal plate lens. The second version XN-2 used a conventional 3.65 meter [12 ft] parabolic antenna, and was the version which went to production. These radars played an important part in the Mercury, Gemini, and early Apollo missions, being deployed in Bermuda, Tannarive, and Australia, among other places for that purpose. The IRACQ [Increased Range ACQuisition] modification was installed on certain of these installations; certainly the one located at Woomera, Australia was so modified. One of the larger installations first appeared in the 1970s as the US Navy's AN/SPY-1 radar used on the Aegis Combat System, and MK-74 radar used on Tartar Guided Missile Fire Control System and research. The cost and complexity of implementing monopulse tracking was reduced and reliability increased when digital signal processing became available after the 1970s. The technology is found in most modern tracking radars and many types of disposable ordnance like missiles.

Passive Radar

Passive radar systems (also referred to as passive coherent location and passive covert radar) encompass a class of radar systems that detect and track objects by processing reflections from non-cooperative sources of illumination in the environment, such as commercial broadcast and communications signals. It is a specific case of bistatic radar, the latter also including the exploitation of cooperative and non-cooperative radar transmitters.

Conventional radar systems comprise a collocated transmitter and receiver, which usually share a common antenna to transmit and receive. A pulsed signal is transmitted and the time taken for the pulse to travel to the object and back allows the range of the object to be determined.

In a passive radar system, there is no dedicated transmitter. Instead, the receiver uses third-party transmitters in the environment, and measures the time difference of arrival between the signal arriving directly from the transmitter and the signal arriving via reflection from the object. This allows the bistatic range of the object to be determined. In addition to bistatic range, a passive radar will typically also measure the bistatic Doppler shift of the echo and also its direction of arrival. These allow the location, heading and speed of the object to be calculated. In some cases, multiple transmitters and/or receivers can be employed to make several independent measurements of bistatic range, Doppler and bearing and hence significantly improve the final track accuracy.

The term "passive radar" is sometimes used incorrectly to describe those passive sensors that detect and track aircraft by their RF emissions (such as radar, communications, or transponder emissions). However, these systems do not exploit reflected energy and hence are more accurately described as ESM systems. Well known examples include the Czech TAMARA and VERA systems and the Ukrainian Kolchuga system.

History

The concept of passive radar detection using reflected ambient radio signals emanating from a distant transmitter is not new. The first radar experiments in the United Kingdom in 1935 by Robert Watson-Watt demonstrated the principle of radar by detecting a Handley Page Heyford bomber at a distance of 12 km using the BBC shortwave transmitter at Daventry.

Early radars were all bistatic because the technology to enable an antenna to be switched from transmit to receive mode had not been developed. Thus many countries were using bistatic systems in air defence networks during the early 1930s. For example, the British deployed the CHAIN HOME system; the French used a bistatic Continuous Wave (CW) radar in a "fence" (or "barrier") system; the Soviet Union deployed a bistatic CW system called the RUS-1; and the Japanese developed a bistatic CW radar called "Type A".

The Germans used a passive bistatic system during World War II. This system, called the Kleine Heidelberg Parasit or Heidelberg-Gerät, was deployed at seven sites (Limmen, Oostvoorne, Ostend, Boulogne, Abbeville, Cap d'Antifer and Cherbourg) and operated as bistatic receivers, using the British Chain Home radars as non-cooperative illuminators, to detect aircraft over the southern part of the North Sea.

Bistatic radar systems gave way to monostatic systems with the development of the synchronizer in 1936. The monostatic systems were much easier to implement since they eliminated the geometric complexities introduced by the separate transmitter and receiver sites. In addition, aircraft and shipborne applications became possible as smaller components were developed. In the early 1950s, bistatic systems were considered again when some interesting properties of the scattered radar energy were discovered, indeed the term "bistatic" was first used by Siegel in 1955 in his report describing these properties.

Experiments in the US led to the deployment of a bistatic system, designated the AN/FPS-23 fluttar radar, in the North American Distant Early Warning (DEW) Line. The fluttar radar was a CW fixed-beam bistatic fence radar developed in 1955 to detect penetration of the DEW line by low-flying bombers. The fluttar radars were designed to fill the low-altitude gaps between SENTINEL monostatic surveillance radars. Fluttar radars were deployed on the DEW line for approximately five years.

The rise of cheap computing power and digital receiver technology in the 1980s led to a resurgence of interest in passive radar technology. For the first time, these allowed designers to apply digital signal processing techniques to exploit a variety of broadcast signals and to use cross-correlation techniques to achieve sufficient signal processing gain to detect targets and estimate their bistatic range and Doppler shift. Classified programmes existed in several nations, but the first announcement of a commercial system was by Lockheed-Martin Mission Systems in 1998, with the commercial launch of the Silent Sentry system, that exploited FM radio and analogue television transmitters.

Typical Illuminators

Passive radar systems have been developed that exploit the following sources of illumination:

- Analog television signals
- FM radio signals
- Cellular phone base stations
- Digital audio broadcasting
- Digital video broadcasting
- Terrestrial High-definition television transmitters in North America
- GPS satellites (GPS reflectometry).

Satellite signals have generally been found to be inadequate for passive radar use: either because the powers are too low, or because the orbits of the satellites are such that illumination is too infrequent. The possible exception to this is the exploitation of satellite-based radar and satellite radio systems. In 2011, researchers Barott and Butka from Embry-Riddle Aeronautical University announced results claiming success using XM Radio to detect aircraft with a low-cost ground station.

Principle

In a conventional radar system, the time of transmission of the pulse and the transmitted waveform are exactly known. This allows the object range to be easily calculated and for a matched filter to be used to achieve an optimal signal-to-noise ratio in the receiver. A passive radar does not have this information directly and hence must use a dedicated receiver channel (known as the "reference channel") to monitor each transmitter being exploited, and dynamically sample the transmitted waveform. A passive radar typically employs the following processing steps:

- Reception of the direct signal from the transmitter(s) and from the surveillance region on dedicated low-noise, linear, digital receivers

- Digital beamforming to determine the direction of arrival of signals and spatial rejection of strong in-band interference
- Adaptive filtering to cancel any unwanted direct signal returns in the surveillance channel(s)
- Transmitter-specific signal conditioning
- Cross-correlation of the reference channel with the surveillance channels to determine object bistatic range and Doppler
- Detection using constant false alarm rate (CFAR) scheme
- Association and tracking of object returns in range/Doppler space, known as "line tracking"
- Association and fusion of line tracks from each transmitter to form the final estimate of an object's location, heading and speed.

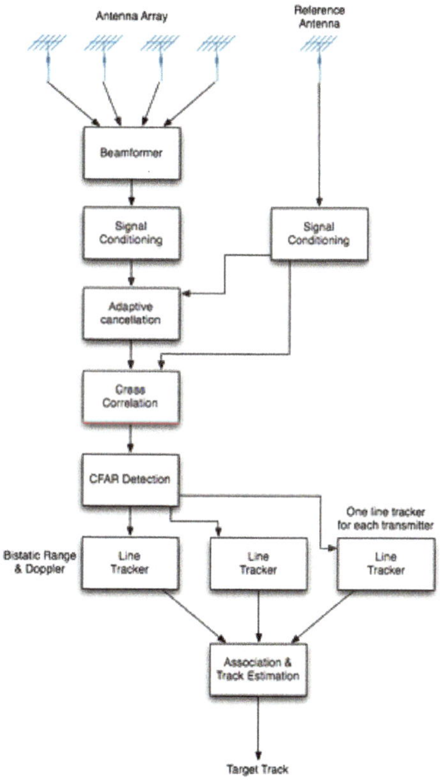

Generic passive radar signal processing scheme

Receiver System

A passive radar system must detect very small target returns in the presence of very strong, continuous interference. This contrasts with a conventional radar, which listens for echoes during the periods of silence in between each pulse transmission. As a result, it is essential that the receiver should have a low noise figure, high dynamic range and high linearity. Despite this, the received echoes are normally well below the noise floor and the system tends to be externally noise limited (due to reception of the transmitted signal itself, plus reception of other distant in-band transmitters). Passive radar systems use digital receiver systems which output a digitized, sampled signal.

Digital Beamforming

Most passive radar systems use simple antenna arrays with several antenna elements and element-level digitisation. This allows the direction of arrival of echoes to be calculated using standard radar beamforming techniques, such as amplitude monopulse using a series of fixed, overlapping beams or more sophisticated adaptive beamforming. Alternatively, some research systems have used only a pair of antenna elements and the phase-difference of arrival to calculate the direction of arrival of the echoes (known as phase interferometry and similar in concept to Very Long Baseline Interferometry used in astronomy).

Signal Conditioning

With some transmitter types, it is necessary to perform some transmitter-specific conditioning of the signal before cross-correlation processing. This may include high quality analogue bandpass filtering of the signal, channel equalization to improve the quality of the reference signal, removal of unwanted structures in digital signals to improve the radar ambiguity function or even complete reconstruction of the reference signal from the received digital signal.

Adaptive Filtering

The principal limitation in detection range for most passive radar systems is the signal-to-interference ratio, due to the large and constant direct signal received from the transmitter. To remove this, an adaptive filter can be used to remove the direct signal in a process similar to active noise control. This step is essential to ensure that the range/Doppler sidelobes of the direct signal do not mask the smaller echoes in the subsequent cross-correlation stage.

In a few specific cases, the direct interference is not a limiting factor, due to the transmitter being beyond the horizon or obscured by terrain (such as with the Manastash Ridge Radar), but this is the exception rather than the rule, as the transmitter must normally be within line-of-sight of the receiver to ensure good low-level coverage.

Cross-correlation Processing

The key processing step in a passive radar is cross-correlation. This step acts as the matched filter and also provides the estimates of the bistatic range and bistatic Doppler shift of each target echo. Most analogue and digital broadcast signals are noise-like in nature, and as a consequence they tend to only correlate with themselves. This presents a problem with moving targets, as the Doppler shift imposed on the echo means that it will not correlate with the direct signal from the transmitter. As a result, the cross-correlation processing must implement a bank of matched filters, each matched to a different target Doppler shift. Efficient implementations of the cross-correlation processing based on the discrete Fourier transform are usually used. The signal processing gain is typically equal to the time-bandwidth product, BT, where B is the waveform bandwidth and T is the length of the signal sequence being integrated. A gain of 50 dB is not uncommon. Extended integration times are limited by the motion of the target and its smearing in range and Doppler during the integration period.

Target Detection

Targets are detected on the cross-correlation surface by applying an adaptive threshold, and declaring all returns above this surface to be targets. A standard cell-averaging constant false alarm rate (CFAR) algorithm is typically used.

Line Tracking

The line-tracking step refers to the tracking of target returns from individual targets, over time, in the range-Doppler space produced by the cross-correlation processing. A standard Kalman filter is typically used. Most false alarms are rejected during this stage of the processing.

Track Association and State Estimation

In a simple bistatic configuration (one transmitter and one receiver) it is possible to determine the location of the target by simply calculating the point of intersection of the bearing with the bistatic-range ellipse. However, errors in bearing and range tend to make this approach fairly inaccurate. A better approach is to estimate the target state (location, heading and speed) from the full measurement set of bistatic range, bearing and Doppler using a non-linear filter, such as the extended or unscented Kalman filter.

When multiple transmitters are used, a target can be potentially detected by every transmitter. The return from this target will appear at a different bistatic range and Doppler shift with each transmitter and so it is necessary to determine which target returns from one transmitter correspond with those on the other transmitters. Having associated these returns, the point at which the bistatic range ellipses from each transmitter intersect is the location of the target. The target can be located much more accurately in this way, than by relying on the intersection of the (inaccurate) bearing measurement with a single range ellipse. Again the optimum approach is to combine the measurements from each transmitter using a non-linear filter, such as the extended or unscented Kalman filter.

Narrow Band and CW Illumination Sources

The above description assumes that the waveform of the transmitter being exploited possesses a usable radar ambiguity function and hence cross-correlation yields a useful result. Some broadcast signals, such as analogue television, contain a structure in the time domain that yields a highly ambiguous or inaccurate result when cross-correlated. In this case, the processing described above is ineffective. If the signal contains a continuous wave (CW) component, however, such as a strong carrier tone, then it is possible to detect and track targets in an alternative way. Over time, moving targets will impose a changing Doppler shift and direction of arrival on the CW tone that is characteristic of the location, speed and heading of the target. It is therefore possible to use a non-linear estimator to estimate the state of the target from the time history of the Doppler and bearing measurements. Work has been published that has demonstrated the feasibility of this approach for tracking aircraft using the vision carrier of analogue television signals. However, track initiation is slow and difficult, and so the use of narrow band signals is probably best considered as an adjunct to the use of illuminators with better ambiguity surfaces.

Performance

Passive radar performance is comparable to conventional short and medium range radar systems. Detection range can be determined using the standard radar equation, but ensuring proper account of the processing gain and external noise limitations is taken. Furthermore, unlike conventional radar, detection range is also a function of the deployment geometry, as the distance of the receiver from the transmitter determines the level of external noise against which the targets must be detected. However, as a rule of thumb it is reasonable to expect a passive radar using FM radio stations to achieve detection ranges of up to 150 km, for high-power analogue TV and US HDTV stations to achieve detection ranges of over 300 km and for lower power digital signals (such as cell phone and DAB or DVB-T) to achieve detection ranges of a few tens of kilometers.

Passive radar accuracy is a strong function of the deployment geometry and the number of receivers and transmitters being used. Systems using only one transmitter and one receiver will tend to be much less accurate than conventional surveillance radars, whilst multistatic radars are capable of achieving somewhat greater accuracies. Most passive radars are two-dimensional, but height measurements are possible when the deployment is such there is significant variation in the altitudes of the transmitters, receiver and target, reducing the effects of geometrical dilution of precision (GDOP).

Advantages and Disadvantages

Advocates of the technology cite the following advantages:

- Lower procurement cost
- Lower costs of operation and maintenance, due to the lack of transmitter and moving parts
- Covert operation, including no need for frequency allocations
- Physically small and hence easily deployed in places where conventional radars cannot be
- Capabilities against stealth aircraft due to the frequency bands and multistatic geometries employed
- Rapid updates, typically once a second
- Difficulty of jamming
- Resilience to anti-radiation missiles.

Opponents of the technology cite the following disadvantages:

- Immaturity
- Reliance on third-party illuminators
- Complexity of deployment
- 1D/2D operation, but possible use 2 *different* systems for 3D (height + range).

Commercial & Academic Systems

Passive radar systems are currently under development in several commercial organizations. Of these, the systems that have been publicly announced include:

- Lockheed-Martin's Silent Sentry - exploiting FM radio stations
- BAE Systems' CELLDAR - exploiting GSM base stations
- Selex ES' Aulos passive radar
- Thales Air Systems' Homeland Alerter - FM radio based system
- Cassidian multiband passive radar
- ERA reports their future VERA-NG may include Passive Radar capabilities.

Several academic passive radar systems exist as well

- Manastash Ridge Radar
- International Project for Radio Meteor Observation
- Alim radar system.

Current Research

Research on passive radar systems is of growing interest throughout the world, with various open source publications showing active research and development in the United States (including work at the Air Force Research Labs, Lockheed-Martin Mission Systems, Raytheon, University of Washington, Georgia Tech/Georgia Tech Research Institute and the University of Illinois), in the NATO C3 Agency in The Netherlands, in the United Kingdom (at Roke Manor Research, QinetiQ, University of Birmingham, University College London and BAE Systems, France (including the government labs of ONERA), Germany (including the labs at Fraunhofer-FHR), Poland (including Warsaw University of Technology). There is also active research on this technology in several government or university laboratories in China, Iran, Russia and South Africa. The low cost nature of the system makes the technology particularly attractive to university laboratories and other agencies with limited budgets, as the key requirements are less hardware and more algorithmic sophistication and computational power.

Much current research is currently focusing on the exploitation of modern digital broadcast signals. The US HDTV standard is particularly good for passive radar, having an excellent ambiguity function and very high power transmitters. The DVB-T digital TV standard (and related DAB digital audio standard) used throughout most of the rest of the world is more challenging—transmitter powers are lower, and many networks are set up in a "single frequency network" mode, in which all transmitters are synchronised in time and frequency. Without careful processing, the net result for a passive radar is like multiple repeater jammers.

At least one non-commercial journalist has suggested that passive radars might likely already be in use worldwide by Police forces and secret services, for legal and illegal wiretapping and eavesdropping.

Target Imaging

Researchers at the University of Illinois at Urbana-Champaign and Georgia Institute of Technology, with the support of DARPA and NATO C3 Agency, have shown that it is possible to build a synthetic aperture image of an aircraft target using passive multistatic radar. Using multiple transmitters at different frequencies and locations, a dense data set in Fourier space can be built for a given target. Reconstructing the image of the target can be accomplished through an inverse fast Fourier transform (IFFT). Herman, Moulin, Ehrman and Lanterman have published reports based on simulated data, which suggest that low frequency passive radars (using FM radio transmissions) could provide target classification in addition to tracking information. These Automatic Target Recognition systems use the power received to estimate the RCS of the target. The RCS estimate at various aspect angles as the target traverses the multistatic system are compared to a library of RCS models of likely targets in order to determine target classification. In the latest work, Ehrman and Lanterman implemented a coordinated flight model to further refine the RCS estimate.

Ionospheric Turbulence Studies

Researchers at the University of Washington operate a distributed passive radar exploiting FM broadcasts to study ionospheric turbulence at altitudes of 100 km and ranges out to 1200 km. Meyer and Sahr have demonstrated interferometric images of ionospheric turbulence with angular resolution of 0.1 degree, while also resolving the full, unaliased Doppler Power Spectrum of the turbulence.

Pulse-Doppler Radar

Airborne pulse-Doppler radar antenna

A pulse-Doppler radar is a radar system that determines the range to a target using pulse-timing techniques, and uses the Doppler effect of the returned signal to determine the target object's velocity. It combines the features of pulse radars and continuous-wave radars, which were formerly separate due to the complexity of the electronics.

Pulse-Doppler systems were first widely used on fighter aircraft starting in the 1960s. Earlier radars had used pulse-timing in order to determine range and the angle of the antenna (or similar means) to determine the bearing. However, this only worked when the radar antenna was not pointed down; in that case the reflection off the ground overwhelmed any returns from other objects. As the ground moves at the same speed but opposite direction of the aircraft, Doppler techniques allow the ground return to be filtered out, revealing aircraft and vehicles. This gives pulse-Doppler radars "look-down/shoot-down" capability. A secondary advantage in military radar is to reduce the transmitted power while achieving acceptable performance for improved safety of stealthy radar.

Pulse-Doppler techniques also find widespread use in meteorological radars, allowing the radar to determine wind speed from the velocity of any precipitation in the air. Pulse-Doppler radar is also the basis of synthetic aperture radar used in radar astronomy, remote sensing and mapping. In air traffic control, they are used for discriminating aircraft from clutter. Besides the above conventional surveillance applications, pulse-Doppler radar has been successfully applied in healthcare, such as fall risk assessment and fall detection, for nursing or clinical purposes.

History

The earliest radar systems failed to operate as expected. The reason was traced to Doppler effects that degrade performance of systems not designed to account for moving objects. Fast-moving objects cause a phase-shift on the transmit pulse that can produce signal cancellation. Doppler has maximum detrimental effect on moving target indicator systems, which must use reverse phase shift for Doppler compensation in the detector.

Doppler weather effects (precipitation) were also found to degrade conventional radar and moving target indicator radar, which can mask aircraft reflections. This phenomenon was adapted for use with weather radar in the 1950s after declassification of some World War II systems.

Pulse-Doppler radar was developed during World War II to overcome limitations by increasing pulse repetition frequency. This required the development of klystron, the traveling wave tube, and solid state devices. Pulse-Doppler is incompatible with other high power microwave amplification devices that are not coherent.

Early examples of military systems include the AN/SPG-51B developed during the 1950s specifically for the purpose of operating in hurricane conditions with no performance degradation.

The Hughes AN/ASG-18 Fire Control System was a prototype airborne radar/combination system for the planned North American XF-108 Rapier interceptor aircraft for the United States Air Force, and later for the Lockheed YF-12. The US's first pulse-Doppler radar, the system had look-down/shoot-down capability and could track one target at a time.

Weather, chaff, terrain, flying techniques, and stealth are common tactics used to hide aircraft from radar. Pulse-Doppler radar eliminates these weaknesses.

It became possible to use pulse-Doppler radar on aircraft after digital computers were incorporated in the design. Pulse-Doppler provided look-down/shoot-down capability to support air-to-air missile systems in most modern military aircraft by the mid 1970s.

Principle

Range Measurement

Principle of pulse-Doppler radar

Principle of pulsed radar

Pulse-Doppler systems measure the range to objects by measuring the elapsed time between sending a pulse of radio energy and receiving a reflection of the object. Radio waves travel at the speed of light, so the distance to the object is the elapsed time multiplied by the speed of light, divided by two - there and back.

Velocity Measurement

Change of wavelength caused by motion of the source

Pulse-Doppler radar is based on the Doppler effect, where movement in range produces frequency shift on the signal reflected from the target.

$$\text{Doppler Frequency} = \left(\frac{2 \times \text{Transmit Frequency} \times \text{Range Velocity}}{C} \right)$$

Radial velocity is essential for pulse-Doppler radar operation. As the reflector moves between each transmit pulse, the returned signal has a phase difference or *phase shift* from pulse to pulse. This causes the reflector to produce Doppler modulation on the reflected signal.

Pulse-Doppler radars exploit this phenomenon to improve performance.

The amplitude of the successively returning pulse from the same scanned volume is:

$$I = I_0 \sin\left(\frac{4\pi(x_0 + v\Delta t)}{\lambda} \right) = I_0 \sin(\Theta_0 + \Delta\Theta)$$

where

$$x_0 = \text{distance radar to target}$$
$$\lambda = \text{radar wavelength}$$
$$\Delta t = \text{time between two pulses}$$

So

$$\Delta\Theta = \left(\frac{4\pi v \Delta t}{\lambda}\right)$$

This allows the radar to separate the reflections from multiple objects located in the same volume of space by separating the objects using a spread spectrum to segregate different signals.

$v = \text{target speed} = \frac{\lambda\Delta\Theta}{4\pi\Delta t}$ where $\Delta\Theta$ is the phase shift induced by range motion.

Benefits

Rejection speed is selectable on pulse-Doppler aircraft-detection systems so nothing below that speed will be detected. A one degree antenna beam illuminates millions of square feet of terrain at 10 miles (16 km) range, and this produces thousands of detections at or below the horizon if Doppler is not used.

Pulse-Doppler radar uses the following signal processing criteria to exclude unwanted signals from slow-moving objects. This is also known as clutter rejection. Rejection velocity is usually set just above the prevailing wind speed (10 to 100 mile/hour or 15 to 150 km/hour). The velocity threshold is much lower for weather radar.

$$\left|\frac{\text{Doppler Frequency} \times C}{2 \times \text{Transmit Frequency}}\right| > \text{Velocity Threshold}$$

In airborne pulse-Doppler radar, the velocity threshold is offset by the speed of the aircraft relative to the ground.

$$\left|\frac{\text{Doppler Frequency} \times C}{2 \times \text{Transmit Frequency}} - \text{Ground Speed} \times \cos(\Theta)\right| > \text{Velocity Threshold}$$

Where Θ is the angle offset between the antenna position and the aircraft flight trajectory.

Surface reflections appear in almost all radar. Ground clutter generally appears in a circular region within a radius of about 25 miles near ground-based radar. This distance extends much further in airborne and space radar. Clutter results from radio energy being reflected from the earth's surface, buildings, and vegetation. Clutter includes weather in radar intended to detect and report aircraft and spacecraft.

Clutter creates a vulnerability region in pulse-amplitude time-domain radar. Non-Doppler radar systems cannot be pointed directly at the ground due to excessive false alarms, which overwhelm computers and operators. Sensitivity must be reduced near clutter to avoid overload. This vulnerability begins in the low-elevation region several beam widths above the horizon, and extends downward. This also exists throughout the volume of moving air associated with weather phenomenon.

Pulse-Doppler radar corrects this as follows.

- Allows the radar antenna to be pointed directly at the ground without overwhelming the computer and without reducing sensitivity.
- Fills in the vulnerability region associated with pulse-amplitude time-domain radar for small object detection near terrain and weather.
- Increases detection range by 300% or more in comparison to moving target indication (MTI) by improving sub-clutter visibility.

Clutter rejection capability of about 60 dB is needed for look-down/shoot-down capability, and pulse-Doppler is the only strategy that can satisfy this requirement. This eliminates vulnerabilities associated with the low-elevation and below-horizon environment.

Pulse compression, and moving target indicator (MTI) provide up to 25 dB sub-clutter visibility. MTI antenna beam is aimed above horizon to avoid excessive false alarm rate, which renders systems vulnerable. Aircraft and some missiles exploit this weakness using a technique called flying below the radar to avoid detection (Nap-of-the-earth). This flying technique is ineffective against pulse-Doppler radar.

Pulse-Doppler provides an advantage when attempting to detect missiles and low observability aircraft flying near terrain, sea surface, and weather.

Audible Doppler and target size support passive vehicle type classification when identification friend or foe is not available from a transponder signal. Medium pulse repetition frequency (PRF) reflected microwave signals fall between 1,500 and 15,000 cycle per second, which is audible. This means a helicopter sounds like a helicopter, a jet sounds like a jet, and propeller aircraft sound like propellers. Aircraft with no moving parts produce a tone. The actual size of the target can be calculated using the audible signal.

Detriments

Ambiguity processing is required when target range is above the red line in the graphic, which increases scan time.

Scan time is a critical factor for some systems because vehicles moving at or above the speed of sound can travel one mile (1.6 km) every few seconds, like the Exocet, Harpoon, Kitchen, and Air-to-air missile. The maximum time to scan the entire volume of the sky must be on the order of a dozen seconds or less for systems operating in that environment.

Pulse-Doppler radar by itself can be too slow to cover the entire volume of space above the horizon unless fan beam is used. This approach is used with the AN/SPS 49(V)5 Very Long Range Air Surveillance Radar, which sacrifices elevation measurement to gain speed.

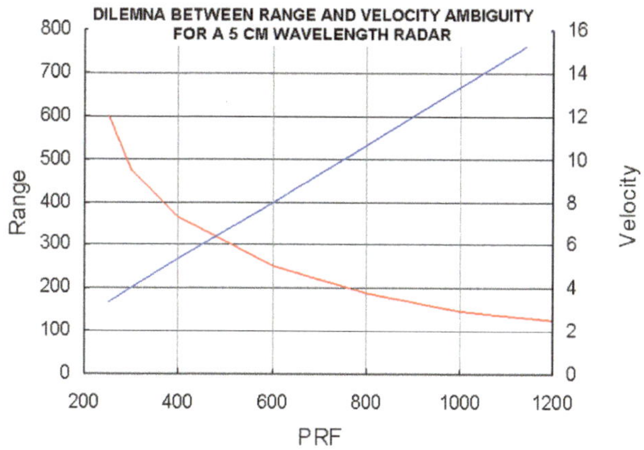

Maximum range from reflectivity (red) and unambiguous Doppler velocity range (blue) with a fix pulse repetition rate.

Pulse-Doppler antenna motion must be slow enough so that all the return signals from at least 3 different PRF can be processed out to the maximum anticipated detection range. This is known as dwell time. Antenna motion for pulse-Doppler must be as slow as radar using MTI.

Search radar that include pulse-Doppler are usually dual mode because best overall performance is achieved when pulse-Doppler is used for areas with high false alarm rates (horizon or below and weather), while conventional radar will scan faster in free-space where false alarm rate is low (above horizon with clear skies).

The antenna type is an important consideration for multi-mode radar because undesirable phase shift introduced by the radar antenna can degrade performance measurements for sub-clutter visibility.

Signal Processing

The signal processing enhancement of pulse-Doppler allows small high-speed objects to be detected in close proximity to large slow moving reflectors. To achieve this, the transmitter must be coherent and should produce low phase noise during the detection interval, and the receiver must have large instantaneous dynamic range.

- Pulse-Doppler signal processing detailed explanation

Pulse-Doppler signal processing also includes ambiguity resolution to identify true range and velocity.

- Ambiguity resolution detailed explanation

The received signals from multiple PRF are compared to determine true range using the range ambiguity resolution process.

- Range ambiguity resolution detailed explanation

The received signals are also compared using the frequency ambiguity resolution process.

- Frequency ambiguity resolution detailed explanation

Range Resolution

The range resolution is the minimum range separation between two objects traveling at the same speed before the radar can detect two discrete reflections.

$$\text{Range Resolution} = \left(\frac{C}{\text{PRF} \times (\text{Number of Samples Between Transmit Pulses})} \right)$$

Velocity Resolution

The velocity resolution is the minimum radial velocity difference between two objects traveling at the same range before the radar can detect two discrete reflections.

$$\text{Velocity Resolution} = \left(\frac{C \times \text{PRF}}{(\text{Transmit Frequency}) \times (\text{Filter Size In Transmit Pulses})} \right)$$

Special Consideration

Pulse-Doppler radar has special requirements that must be satisfied to achieve acceptable performance.

Pulse Repetition Frequency

Pulse-Doppler typically uses medium pulse repetition frequency (PRF) from about 3 kHz to 30 kHz. The range between transmit pulses is 5 km to 50 km.

Range and velocity cannot be measured directly using medium PRF, and ambiguity resolution is required to identify true range and speed. Doppler signals are generally above 1 kHz, which is audible, so audio signals from medium-PRF systems can be used for passive target classification.

Angular Measurement

Radar systems require angular measurement. Transponders are not normally associated with pulse-Doppler radar, so sidelobe suppression is required for practical operation.

Tracking radar systems use angle error to improve accuracy by producing measurements perpendicular to the radar antenna beam. Angular measurements are averaged over a span of time and combined with radial movement to develop information suitable to predict target position for a short time into the future.

The two angle error techniques used with tracking radar are monopulse and conical scan.

- Monopulse radar
- Conical scanning

Coherency

Pulse-Doppler radar requires a coherent oscillator with very little noise. Phase noise reduces sub-clutter visibility performance by producing apparent motion on stationary objects.

Cavity magnetron and crossed-field amplifier are not appropriate because noise introduced by these devices interfere with detection performance. The only amplification devices suitable for pulse-Doppler are klystron, traveling wave tube, and solid state devices.

Scalloping

Pulse-Doppler signal processing introduces a phenomenon called scalloping. The name is associated with a series of holes that are scooped-out of the detection performance.

Scalloping for pulse-Doppler radar involves blind velocities created by the clutter rejection filter. Every volume of space must be scanned using 3 or more different PRF. A two PRF detection scheme will have detection gaps with a pattern of discrete ranges, each of which has a blind velocity.

Windowing

Ringing artifacts pose a problem with search, detection, and ambiguity resolution in pulse-Doppler radar.

Ringing is reduced in two ways.

First, the shape of the transmit pulse is adjusted to smooth the leading edge and trailing edge so that RF power is increased and decreased without an abrupt change. This creates a transmit pulse with smooth ends instead of a square wave, which reduces ringing phenomenon that is otherwise associated with target reflection.

Second, the shape of the receive pulse is adjusted using a window function that minimizes ringing that occurs any time pulses are applied to a filter. In a digital system, this adjusts the phase and/or amplitude of each sample before it is applied to the Fast Fourier Transform. The Dolph-Chebyshev window is the most effective because it produces a flat processing floor with no ringing that would otherwise cause false alarms.

Antenna

Pulse-Doppler radar is generally limited to mechanically aimed antennas and active phase array.

Mechanical RF components, such as wave-guide, can produce Doppler modulation due to phase shift induced by vibration. This introduces a requirement to perform full spectrum operational tests using shake tables that can produce high power mechanical vibration across all anticipated audio frequencies.

Doppler is incompatible with most electronically steered phase-array antenna. This is because the phase-shifter elements in the antenna are non-reciprocal and the phase shift must be adjusted before and after each transmit pulse. Spurious phase shift is produced by the sudden impulse of the phase shift, and settling during the receive period between transmit pulses places Doppler modulation onto stationary clutter. That receive modulation corrupts the measure of performance for sub-clutter visibility. Phase shifter settling time on the order of 50ns is required. Start of receiver sampling needs to be postponed at least 1 phase-shifter settling time-constant (or more) for each 20 dB of sub-clutter visibility.

Most antenna phase shifters operating at PRF above 1 kHz introduce spurious phase shift unless special provisions are made, such as reducing phase shifter settling time to a few dozen nanoseconds.

The following gives the maximum permissible settling time for antenna phase shift modules.

$$T = \frac{1}{e^{\frac{SCV}{20}} \times S \times PRF}$$

where

- T = phase shifter settling time
- SCV = sub-clutter visibility in dB
- S = number of range samples between each transmit pulse
- PRF = maximum design pulse repetition frequency

The antenna type and scan performance is a practical consideration for multi-mode radar systems.

Diffraction

Choppy surfaces, like waves and trees, form a diffraction grating suitable for bending microwave signals. Pulse-Doppler can be so sensitive that diffraction from mountains, buildings or wave tops can be used to detect fast moving objects otherwise blocked by solid obstruction along the line of sight. This is a very lossy phenomenon that only becomes possible when radar has significant excess sub-clutter visibility.

Refraction and ducting use transmit frequency at L-band or lower to extend the horizon, which is very different from diffraction. Refraction for over-the-horizon radar uses variable density in the air column above the surface of the earth to bend RF signals. An inversion layer can produce a transient troposphere duct that traps RF signals in a thin layer of air like a wave-guide.

Subclutter Visibility

Subclutter visibility involves the maximum ratio of clutter power to target power, which is proportional to dynamic range. This determines performance in heavy weather and near the earth surface.

$$\text{Dynamic Range} = \text{Min}\begin{cases} \frac{\text{Carrier Power}}{\text{Noise Power}} & \text{Transmit Noise, where bandwidth is} \frac{\text{PRF}}{\text{Filter Size}} \\ 2^{(\text{Sample Bits} + \text{Filter Size})} & \text{Receiver Dynamic Range} \end{cases}$$

Subclutter visibility is the ratio of the smallest signal that can be detected in the presence of a larger signal.

$$\text{Subclutter Visibility} = \left(\frac{\text{Dynamic Range}}{\text{CFAR Detection Threshold}}\right)$$

A small fast-moving target reflection can be detected in the presence of larger slow-moving clutter reflections when the following is true.

$$\text{Target Power} > \left(\frac{\text{Clutter Power}}{\text{Subclutter Visibility}}\right)$$

Performance

The pulse-Doppler radar equation can be used to understand trade-offs between different design constraints, like power consumption, detection range, and microwave safety hazards. This is a very simple form of modeling that allows performance to be evaluated in a sterile environment.

The theoretical range performance is as follows.

$$R = \left(\frac{P_t\, G_t\, A_r\, \sigma F\, D}{16\,\pi^2\, K_b\, T\, B\, N}\right)^{\frac{1}{4}}$$

where

- R = Distance to the target
- P_t = Transmitter power
- G_t = Gain of the transmitting antenna
- A_r = Effective aperture (area) of the receiving antenna
- σ = Radar cross section, or scattering coefficient, of the target
- F = Antenna pattern propagation factor
- D = Doppler filter size (transmit pulses in each Fast Fourier transform)
- K_b = Boltzmann's constant
- T = Temperature (kelvin)
- B = Receiver Bandwidth (band pass filter)
- N = Noise figure

This equation is derived by combining the Radar equation with the Noise equation and accounting for in-band noise distribution across multiple detection filters. The value D is added to the standard radar range equation to account for both pulse-Doppler signal processing and transmitter FM noise reduction.

Detection range is increased proportional to the fourth root of the number of filters for a given power consumption. Alternatively, power consumption is reduced by the number of filers for a given detection range.

Pulse-Doppler signal processing integrates all of the energy from all of the individual reflected pulses that enter the filter. This means a pulse-Doppler signal processing system with 1,024 elements provides 30.103 dB of improvement due to the type of signal processing that must be used with pulse-Doppler radar. The energy of all of the individual pulses from the object are added together by the filtering process.

Signal processing for a 1,024 point filter improves performance by 30.103 dB, assuming compatible transmitter and antenna. This corresponds to the following potential improvements.

- 562% increase in maximum distance.

These improvements are the reason pulse-Doppler is essential for military and astronomy.

Aircraft Tracking uses

Pulse-Doppler radar for aircraft detection has two modes.

- Scan
- Track

Scan mode involves frequency filtering, amplitude thresholding, and ambiguity resolution. Once a reflection has been detected and resolved, the pulse-Doppler radar automatically transitions to tracking mode for the volume of space surrounding the track.

Track mode works like a phase-locked loop, where Doppler velocity is compared with the range movement on successive scans. Lock indicates the difference between the two measurements is below a threshold, which can only occur with an object that satisfies Newtonian mechanics. Other types of electronic signals cannot produce a lock. Lock exists in no other type of radar.

The lock criteria needs to be satisfied during normal operation.

- Pulse-Doppler signal processing Lock

Lock eliminates the need for human intervention with the exception of helicopters and electronic jamming.

Weather phenomenon obey adiabatic process associated with air mass and not Newtonian mechanics, so the lock criteria is not normally used for weather radar.

Pulse-Doppler signal processing selectively excludes low-velocity reflections so that no detections occurs below a threshold velocity. This eliminates terrain, weather, biologicals, and mechanical jamming with the exception of decoy aircraft.

The target Doppler signal from the detection is converted from frequency domain back into time domain sound for the operator in track mode on some radar systems. The operator uses this sound for passive target classification, such as recognizing helicopters and electronic jamming.

Helicopters

Special consideration is required for aircraft with large moving parts because pulse-Doppler radar operates like a phase-locked loop. Blade tips moving near the speed of sound produce the only signal that can be detected when a helicopter is moving slow near terrain and weather.

Helicopters appears like a rapidly pulsing noise emitter except in a clear environment free from clutter. An audible signal is produced for passive identification of the type of airborne object. Microwave Doppler frequency shift produced by reflector motion falls into the audible sound range for human beings (20 – 20,000 Hz), which is used for target classification in addition to the kinds of conventional radar display used for that purpose, like A-scope, B-scope, C-scope, and RHI indicator. The human ear may be able to tell the difference better than electronic equipment.

A special mode is required because the Doppler velocity feedback information must be unlinked from radial movement so that the system can transition from scan to track with no lock.

Similar techniques are required to develop track information for jamming signals and interference that cannot satisfy the lock criteria.

Multi-mode

Pulse-Doppler radar must be multi-mode to handle aircraft turning and crossing trajectory.

Once in track mode, pulse-Doppler radar must include a way to modify Doppler filtering for the volume of space surrounding a track when radial velocity falls below the minimum detection velocity. Doppler filter adjustment must be linked with a radar track function to automatically adjust Doppler rejection speed within the volume of space surrounding the track.

Tracking will cease without this feature because the target signal will otherwise be rejected by the Doppler filter when radial velocity approaches zero.

Multi-mode operation may also include continuous wave illumination for semi-active radar homing.

Continuous-wave Radar

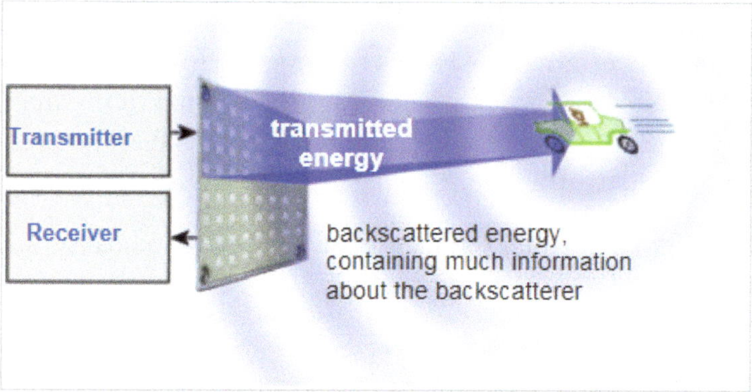

Principle of a measurement with a continuous wave radar

Continuous-wave radar is a type of radar system where a known stable frequency continuous-wave radio energy is transmitted and then received from any reflecting objects. Continuous-wave (CW) radar uses Doppler, which renders the radar immune to interference from large stationary objects and slow moving clutter.

CW radar systems are used at both ends of the range spectrum.

- Inexpensive radio-altimeters, proximity sensors and sport accessories that operate from a few dozen feet to several kilometers.
- Costly early-warning CW angle track (CWAT) radar operating beyond 100 km for use with surface-to-air missile systems.

Operation

The main advantage of CW radar is that energy is not pulsed so these are much simpler to manufacture and operate. They have no minimum or maximum range, although the broadcast power level imposes a practical limit on range. Continuous-wave radar maximize total power on a target because the transmitter is broadcasting continuously.

The military uses continuous-wave radar to guide semi-active radar homing (SARH) air-to-air missiles, such as the U.S. AIM-7 Sparrow and standard missile. The launch aircraft *illuminates* the target with a CW radar signal, and the missile homes in on the reflected radio waves. Since the missile is moving at high velocities relative to the aircraft, there is a strong Doppler shift. Most modern air combat radars, even pulse Doppler sets, have a CW function for missile guidance purposes.

Maximum distance in a continuous-wave radar is determined by the overall bandwidth and transmitter power. This bandwidth is determined by two factors.

- Transmit energy density (watts per Hertz)
- Receiver filter size (bandwidth divided by the total number of filters)

Doubling transmit power increases distance performance by about 20%. Reducing the total FM transmit noise by half has the same effect.

Frequency domain receivers used for continuous-wave Doppler radar receivers are very different from conventional radar receivers. The receiver consists of a bank of filters, usually more than 100. The number of filters determines the maximum distance performance.

Doubling the number of receiver filters increases distance performance by about 20%. Maximum distance performance is achieved when receiver filter size is equal to the maximum FM noise riding on the transmit signal. Reducing receiver filter size below average amount of FM transmit noise will not improve range performance.

A CW radar is said to be *matched* when the receiver filter size matches the RMS bandwidth of the FM noise on the transmit signal.

Types

There are two types of continuous-wave radar: *unmodulated continuous-wave* and *modulated continuous-wave*.

Unmodulated Continuous-wave

This kind of radar can cost less than $100 (2012). Return frequencies are shifted away from the transmitted frequency based on the Doppler effect when objects are moving. There is no way to evaluate distance. This type of radar is typically used with competition sports, like golf, tennis, baseball, and NASCAR racing.

The Doppler frequency change depends on the speed of light in the air (c' is slightly slower than in vacuum) and v the speed of the target:

$$f_r = f_t\left(\frac{1+v/c'}{1-v/c'}\right)$$

The Doppler frequency is thus:

$$f_d = f_r - f_t = 2v\frac{f_t}{(c'-v)}$$

Since the usual variation of targets' speed of a radar is much smaller than $c', (v \ll c')$, it is possible to simplifiy with $(c'-v) \rightarrow c'$:

$$f_d \approx 2v\frac{f_t}{c'}$$

Continuous-wave radar without frequency modulation (FM) only detects moving targets, as stationary targets (along the line of sight) will not cause a Doppler shift. Reflected signals from stationary and slow-moving objects are masked by the transmit signal, which overwhelms reflections from slow-moving objects during normal operation.

Modulated Continuous-wave

Frequency-modulated continuous-wave radar (FM-CW) – also called continuous-wave frequency-modulated (CWFM) radar – is a short-range measuring radar set capable of determining distance. This increases reliability by providing distance measurement along with speed measurement, which is essential when there is more than one source of reflection arriving at the radar antenna. This kind of radar is often used as "radar altimeter" to measure the exact height during the landing procedure of aircraft. It is also used as early-warning radar, wave radar, and proximity sensors. Doppler shift is not always required for detection when FM is used.

In this system the transmitted signal of a known stable frequency continuous wave varies up and down in frequency over a fixed period of time by a modulating signal. Frequency difference between the receive signal and the transmit signal increases with delay, and hence with distance. This smears out, or blurs, the Doppler signal. Echoes from a target are then mixed with the transmitted signal to produce a beat signal which will give the distance of the target after demodulation.

A variety of modulations is possible, the transmitter frequency can slew up and down as follows :

- Sine wave, like air raid siren
- Sawtooth wave, like the chirp from a bird

- Triangle wave, like police siren in the United States
- Square wave, like police siren in the United Kingdom

Range demodulation is limited to 1/4 wavelength of the transmit modulation. Instrumented range for 100 Hz FM would be 500 km. That limit depends upon the type of modulation and demodulation. The following generally applies.

$$\text{Instrumented Range} = F_r - F_t = \frac{\text{Speed of Light}}{(4 \times \text{Modulation Frequency})}$$

The radar will report incorrect distance for reflections from distances beyond the instrumented range, such as from the moon. FMCW range measurements are only reliable to about 60% of the instrumented range, or about 300 km for 100 Hz FM.

Sawtooth Frequency Modulation

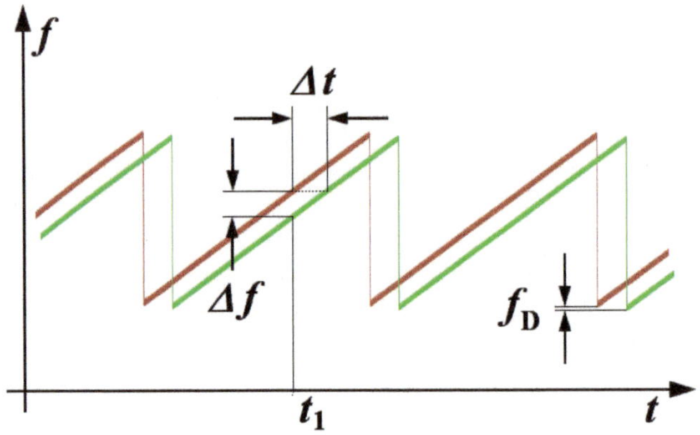

Ranging with an FM-CW radar system: if the error caused by a possible Doppler frequency f_D can be ignored and the transmitter's power is linearly frequency modulated, then the time delay (Δt) is proportional to the difference of the transmitted and the received signal (Δf) at any time.

Sawtooth modulation is the most used in FM-CW radars where range is desired for objects that lack rotating parts. Range information is mixed with the Doppler velocity using this technique. Modulation can be turned off on alternate scans to identify velocity using unmodulated carrier frequency shift. This allows range and velocity to be found with one radar set. Triangle wave modulation can be used to achieve the same goal.

As shown in the figure the received waveform (green) is simply a delayed replica of the transmitted waveform (red). The transmitted frequency is used to down-convert the receive signal to baseband, and the amount of frequency shift between the transmit signal and the reflected signal increases with time delay (distance). The time delay is thus a measure of the range; a small frequency spread is produced by nearby reflections, a larger frequency spread corresponds with more time delay and a longer range.

With the advent of modern electronics, digital signal processing is used for most detection processing. The beat signals are passed through an analog-to-digital converter, and digital processing is performed on the result. As explained in the literature, FM-CW ranging for a linear ramp waveform is given in the following set of equations:

$$k = \frac{\Delta f_{radar}}{\Delta t_{radar}}$$

where Δf_{radar} is the radar frequency sweep amount and Δt_{radar} is the time to complete the frequency sweep.

Then, $\Delta f_{echo} = t_r k$, rearrange to a more useful:

$t_r = \frac{\Delta f_{echo}}{k}$, where t_r is the round trip time of the radar energy.

It is then a trivial matter to calculate the physical one-way distance for an idealized typical case as:

$$\text{dist}_{oneway} = \frac{c' t_r}{2}$$

where $c' = c / n$ is the speed of light in any transparent medium of refractive index n (n=1 in vacuum and 1.0003 for air).

For practical reasons, receive samples are not processed for a brief period after the modulation ramp begins because incoming reflections will have modulation from the previous modulation cycle. This imposes a range limit and limits performance.

$$\text{Range Limit} = 0.5\, c'\, t_{radar}$$

Sinusoidal Frequency Modulation

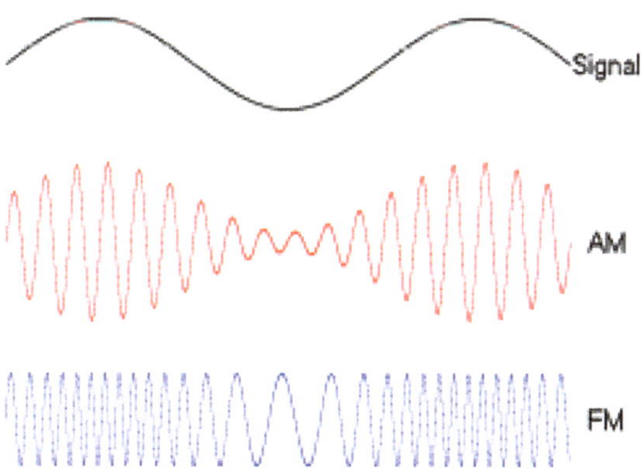

Sinusoidal FM modulation identifies range by measuring the amount of spectrum spread produced by propagation delay (AM is not used with FMCW).

Sinusoidal FM is used when both range and velocity are required simultaneously for complex objects with multiple moving parts like turbine fan blades, helicopter blades, or propellers. This processing reduces the effect of complex spectra modulation produced by rotating parts that introduce errors into range measurement process.

This technique also has the advantage that the receiver never needs to stop processing incoming signals because the modulation waveform is continuous with no impulse modulation.

Sinusoidal FM is eliminated completely by the receiver for close in reflections because the transmit frequency will be the same as the frequency being reflected back into the receiver. The spectrum for more distant objects will contain more modulation. The amount of spectrum spreading caused by modulation riding on the receive signal is proportional to the distance to the reflecting object.

The time domain formula for FM is:

$$y(t) = \cos\left\{2\pi[f_c + \mathrm{B}\cos\left(2\pi f_m t\right)]t\right\}$$

where $\mathrm{B} = \dfrac{f_\Delta}{f_m}$ (modulation index)

A time delay is introduced in transit between the radar and the reflector.

$$y(t) = \cos\left\{2\pi[f_c + \mathrm{B}\cos\left(2\pi f_m (t+\delta t)\right)](t+\delta t)\right\}$$

where $\delta t =$ time delay

The detection process down converts the receive signal using the transmit signal. This eliminates the carrier.

$$y(t) \approx \cos\left\{-4t\pi \mathrm{B}\sin(2\pi f_m (2t+\delta t)\sin(\pi f_m \delta t) + 2\delta t\pi \mathrm{B}\cos(2\pi f_m (t+\delta t))\right\}$$

$$y(t) \approx \cos\left\{-4t\pi \mathrm{B}\sin(2\pi f_m (2t+\delta t)\sin(\pi f_m \delta t) + 2\delta t\pi \mathrm{B}\cos(2\pi f_m (t+\delta t))\right\}$$

The Carson bandwidth rule can be seen in this equation, and that is a close approximation to identify the amount of spread placed on the receive spectrum:

$$\text{Modulation Spectrum Spread} \approx 2(\mathrm{B}+1) f_m \sin(\delta t)$$

$$\text{Range} = 0.5C / \delta t$$

Receiver demodulation is used with FMCW similar to the receiver demodulation strategy used with pulse compression. This takes place before Doppler CFAR detection processing. A large modulation index is needed for practical reasons.

Practical systems introduce reverse FM on the receive signal using digital signal processing before the Fast Fourier Transform process is used to produce the spectrum. This is repeated with several different demodulation values. Range is found by identifying the receive spectrum where width is minimum.

Practical systems also process receive samples for several cycles of the FM in order to reduce the influence of sampling artifacts.

Configurations

There are two different antenna configurations used with continuous-wave radar: *monostatic radar*, and *bistatic radar*.

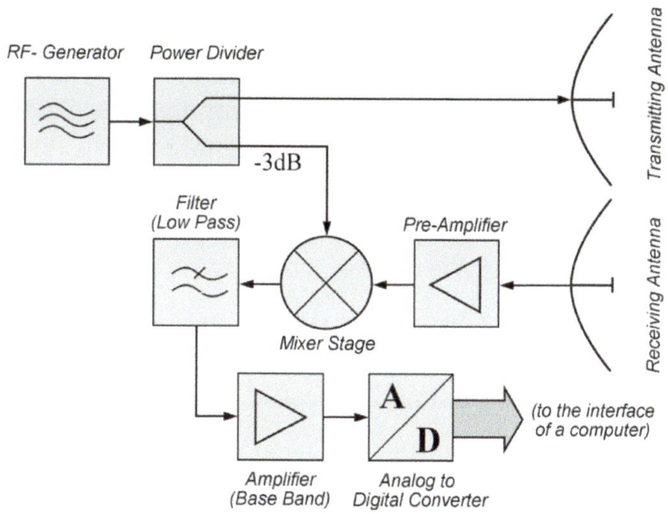

Block diagram of a simple continuous-wave radar module: Many manufacturers offer such transceiver modules and rename them as "Doppler radar sensors"

Monostatic

The radar receive antenna is located nearby the radar transmit antenna in monostatic radar.

Feed-through null is typically required to eliminate bleed-through between the transmitter and receiver to increase sensitivity in practical systems. This is typically used with continuous-wave angle tracking (CWAT) radar receivers that are interoperable with surface-to-air missile systems.

Interrupted continuous-wave can be used to eliminate bleed-through between the transmit and receive antenna. This kind of system typically takes one sample between each pair of transmit pulses, and the sample rate is typically 30 kHz or more. This technique is used with the least expensive kinds of radar, such as those used for traffic monitoring and sports.

FM-CW radars can be built with one antenna using either a circulator, or circular polarization.

Bistatic

The radar receive antenna is located far from the radar transmit antenna in bistatic radar. The transmitter is fairly expensive, while the receiver is fairly inexpensive and disposable.

This is typically used with semi-active radar homing including most surface-to-air missile systems. The transmit radar is typically located near the missile launcher. The receiver is located in the missile.

The transmit antenna illuminates the target in much the same way as a search light. The transmit antenna also issues an omnidirectional sample.

The receiver uses two antennas – one antenna aimed at the target and one antenna aimed at the transmit antenna. The receive antenna that is aimed at the transmit antenna is used to develop the feed-through null, which allows the target receiver to operate reliably in or near the main beam of the antenna.

The bistatic FM-CW receiver and transmitter pair may also take the form of an over-the-air de-ramping (OTAD) system. An OTAD transmitter broadcasts an FM-CW signal on two different frequency channels; one for synchronisation of the receiver with the transmitter, the other for illuminating the measurement scene. Using directive antennas, the OTAD receiver collects both signals simultaneously and mixes the synchronisation signal with the downconverted echo signal from the measurement scene in a process known as over-the-air deramping. The frequency of de-ramped signal is proportional to the bistatic range to the target less the baseline distance between the OTAD transmitter and the OTAD receiver.

Most modern systems FM-CW radars use one transmitter antenna and multiple receiver antennas. Because the transmitter is on continuously at effectively the same frequency as the receiver, special care must be exercised to avoid overloading the receiver stages.

Monopulse

Monopulse antennas produce angular measurements without pulses or other modulation. This technique is used in semi-active radar homing.

Leakage

The transmit signal will leak into the receiver on practical systems. Significant leakage will come from nearby environmental reflections even if antenna components are perfect. As much as 120 dB of leakage rejection is required to achieve acceptable performance.

Three approaches can be used to produce a practical system that will function correctly.

- Null
- Filter
- Interruption

Null and filter approaches must be used with bistatic radar, like semi-active radar homing, for practical reasons because side-lobes from the illumination radar will illuminate the environment in addition to the main-lobe illumination on the target. Similar constraints apply to ground based CW radar. This adds cost.

Interruption applies to cheap hand held mono-static radar systems (police radar and sporting goods). This is impractical for bistatic systems because of the cost and complexity associated with coordinating time with nuclear precision in two different locations.

The design constraint that drives this requirement is the dynamic range limitation of practical receiver components that include band pass filters that take time to settle out.

Null

The null approach takes two signals:

- A sample of the transmit signal leaking into the receiver
- A sample of the actual transmit signal

The actual transmit signal is rotated 180 degrees, attenuated, and fed into the receiver. The phase shift and attenuation are set using feedback obtained from the receiver to cancel most of the leakage. Typical improvement is on the order of 30 dB to 70 dB.

Filter

The filter approach relies on using a very narrow band reject filter that will eliminate low velocity signals from nearby reflectors. The band reject area spans 10 mile per hour to 100 mile per hour depending upon the anticipated environment. Typical improvement is on the order of 30 dB to 70 dB.

Interruption

While interrupted carrier systems are not considered to be CW systems, performance characteristics are sufficiently similar to group interrupted CW systems with pure CW radar because the pulse rate is high enough that range measurements cannot be done without frequency modulation (FM).

This technique turns the transmitter off for a period before receiver sampling begins. Receiver interference declines by about 8.7 dB per time constant. Leakage reduction of 120 dB requires 14 recover bandwidth time constants between when the transmitter is turned off and receiver sampling begins.

Advantages

Because of simplicity, CW radar are inexpensive to manufacture, relatively free from failure, cheap to maintain, and fully automated. Some are small enough to carry in a pocket. More sophisticated CW radar systems can reliably achieve accurate detections exceeding 100 km distance while providing missile illumination.

The FMCW ramp can be compressed providing extra signal to noise gains such one does not need the extra power that pulse radar using a no FM modulation would. This combined with the fact that it is coherent means that Fourier integration can be used rather than azimuth integration providing superior signal to noise and a Doppler measurement.

Doppler processing allows signal integration between successive receiver samples. This means that the number of samples can be increased to extend the detection range without increasing transmit power. That technique can be used to produce inexpensive stealthy low-power radar.

CW performance is similar to Pulse-Doppler radar performance for this reason.

Limitations

Unmodulated continuous wave radar cannot measure distance, and the beam is usually broad with side-lobes that extend to the side and behind the radar antenna. Signal amplitude provides the only way to determine which object corresponds with which speed measurement when there is more than one moving object near the receiver, but amplitude information is not useful without range measurement to evaluate target size. Moving objects include birds flying near objects in front of the antenna. Reflections from small objects directly in front of the receiver can be overwhelmed by reflections entering antenna side-lobes from large object located to the side, above, or behind the radar, such as trees with wind blowing through the leaves, tall grass, sea surface, freight trains, busses, trucks, and aircraft.

Small radar systems that lack range modulation are only reliable when used with one object in a sterile environment free from vegetation, aircraft, birds, weather phenomenon, and other nearby vehicles.

With 20 dB antenna side-lobes, a truck or tree with 1,000 square feet of reflecting surface behind the antenna can produce a signal as strong as a car with 10 square feet of reflecting in front of a small hand held antenna. An area survey is required to determine if hand held devices will operate reliably because unobserved roadway traffic and trees behind the operator can interfere with observations made in front of the operator.

This is a typical problem with radar speed guns used by law enforcement officers, NASCAR events, and sports, like baseball, golf, and tennis. Interference from a second radar, automobile ignition, other moving objects, moving fan blades on the intended target, and other radio frequency sources will corrupt measurements. These systems are limited by wavelength, which is 0.03 meter at Ku band, so the beam spread exceeds 45 degrees if the antenna is smaller than 12 inches (0.3 meter). Significant antenna side-lobes extend in all directions unless the antenna is larger than the vehicle on which the radar is mounted.

Side-lobe suppression and FM range modulation are required for reliable operation. There is no way to know the direction of the arriving signal without side-lobe suppression, which requires two or more antennae, each with its own individual receiver. There is no way to know distance without FM range modulation.

Speed, direction, and distance are all required to pick out an individual object.

These limitations are due to the well known limitations of basic physics that cannot be overcome by design.

Law enforcement agencies include hand held laser in the mix of tools needed for law enforcement to confirm reliable speed and position of an individual vehicle in traffic after radar detects excessive speed.

Space-based Radar

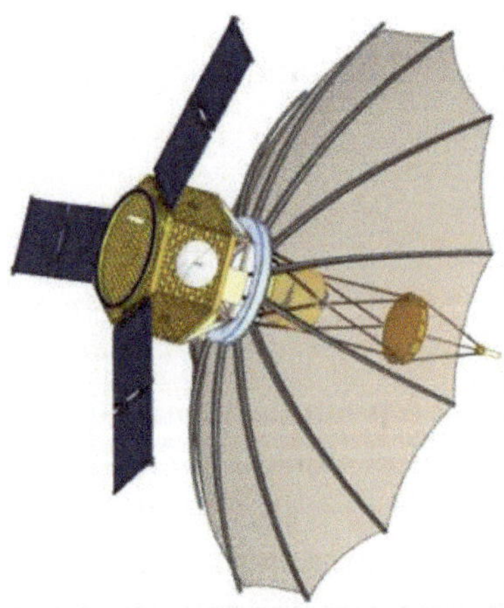

ORS-2

Space-based radar refers to space-borne radar systems that may have any of a variety of purposes. A number of earth-observing radar satellites, such as RADARSAT, have employed synthetic aperture radar (SAR) to obtain terrain and land-cover information about the Earth.

Military

- Lacrosse (satellite)
- SAR-Lupe

Discoverer II was a proposed military space-based radar program initiated in February 1998 as a joint Air Force, DARPA, and NRO program. The concept was to provide high-range-resolution ground moving target indication (GMTI), as well as SAR imaging and high-resolution digital mapping. This program was cancelled by Congress in 2007. SBR is a less-ambitious version of Discoverer II.

Space-based radar (SBR) is a proposed constellation of active radar satellites for the United States Department of Defense. The SBR system would allow detection and tracking of aircraft, ocean-going vessels (similar to the Soviet US-A program), and potentially land vehicles from space. This information would then be relayed to regional and national command centers, as well as E-10 MC2A airborne command posts.

Earth-observing Radars

Use of radar sensor for Earth observation purposes was started by NASA/JPL's Seasat satellite, which carried three different radar sensors:

- a synthetic aperture radar (SAR) for high-resolution imaging
- a radar altimeter, to measure the ocean topography
- a wind scatterometer to measure wind speed and direction

After Seasat, SARs, altimeters and scatterometers have been flown on several other space missions.

While the SAR, in principle, is similar to its airborne counterparts (with the advantage of the increased coverage and worldwide access offered by the satellite platform), the other two are specific to satellite operations.

A satellite radar-altimeter is a nadir-looking radar with very high range resolution, which measures the ocean surface topography with an accuracy in the order of few centimeters. Additionally, analysis of the echo amplitude and shape can extract information about the wind speed and wave height, respectively. Some radar-altimeters (like CryoSat/SIRAL) employ synthetic aperture and/or interferometric techniques: their reduced footprint allows mapping of rougher surfaces like polar ices.

A wind scatterometer observes the same portion of the ocean surface from different (at least 3) angles of view as the satellite passes by, measuring the echo amplitude and the corresponding surface reflectivity. Reflectivity being affected by the ocean surface "roughness", which in turn is affected by the wind and also dependent on its direction, this instrument can determine the wind speed and direction.

These three types of radar are currently used on several satellites. Scatterometers are of high value for operational meteorology, allowing reconstruction of wind fields on a global scale. Data from radar altimeters are used for the accurate determination of the geoid, monitoring of tides, ocean currents and other large-scale ocean phenomena such as El Niño.

SARs applications are many: they range from geology to crop monitoring, from measurement of sea ice to disaster monitoring to vessel traffic surveillance, not to forget the military applications (many civilian SAR satellites are, in fact, dual-use systems). SAR imaging offer the great advantage, over its optical counterparts, of not being affected by meteorological conditions such as clouds, fog, etc., making it the sensor of choice when continuity of data must be ensured. Additionally, SAR interferometry (both dual-pass or single-pass, as used in the SRTM mission) allows accurate 3-D Reconstruction.

Other types of radars have been flown for earth observation missions: precipitation radars such as the Tropical Rainfall Measuring Mission, or cloud radars like the one used on Cloudsat.

Like other Earth observation satellites, radar satellites often use sun-synchronous orbits so that diurnal variations of vegetation are ignored, allowing long-term variations to be more accurately measured.

Earth-observing radar satellites include:

- RISAT-1 (SAR,ISRO India, 2012)
- RORSAT (SAR, Soviet Union, 1967-1988)
- Seasat (SAR, altimeter, scatterometer, US, 1978)
- RADARSAT-1 (SAR, Canadian, 1995)
- RADARSAT-2 (SAR, Canadian, 2007)
- SAR Lupe 1-5 (SAR satellites of the German Air Force)
- TerraSAR-X (SAR Germany, 2007)
- TanDEM-X (SAR Germany, 2010)
- COSMO-SkyMed (SAR, Italy, 2007)
- TecSAR (SAR, Israeli, 2008)
- TOPEX/Poseidon (altimeter)
- Jason 1 / Jason 2 (altimeter)
- Shuttle Imaging Radar
- JERS-1 (SAR)
- Geosat (altimeter)
- ERS-1 & ERS-2 (European Remote-Sensing Satellite) (altimeter, combined SAR/scatterometer)

- Envisat (SAR, altimeter)
- Tropical Rainfall Measuring Mission (Precipitation Radar)
- Cloudsat (cloud radar)
- Metop (scatterometer)
- QuickScat (scatterometer)
- NISAR (satellite)
- Almaz
- Sentinel-1

Planetary Radars

Most of the radars flown as payload in planetary missions (i.e., not considering avionics radar, such as docking and landing radars used in Apollo and LEM) belong to two categories: imaging radars and sounders.

Imaging radars: Synthetic aperture radars are the only instruments capable of penetrating heavy cloud cover around planets such as Venus, which was the first target for such missions. Two Soviet spacecraft (Venera 15 and Venera 16) imaged the planet in 1983 and 1984 using SAR and Radar altimeters. The Magellan probe also imaged Venus in 1990 and 1994.

The only other target of an imaging radar mission has been Titan, the largest moon of Saturn, in order to penetrate its opaque atmosphere. The radar of the Cassini probe, in orbit around Saturn, is currently providing images of Titan surface at each fly-by of the moon. The Cassini radar is a multimode system and can operate as Synthetic Aperture Radar, radar altimeter, scatterometer and radiometer.

Sounding radars: these are low-frequency (normally, HF - 3 to 30 MHz - or lower) ground-penetrating Radars, used to acquire data about the planet sub-surface structure. Their low operating frequency allow them to penetrate hundreds of meters, or even kilometers, below the surface. Synthetic aperture techniques are normally exploited to reduce the ground footprint (due to the low operating frequency and the small allowable antenna dimensions, the beam is very wide) and, thus, the unwanted echo from other surface objects.

The first radar sounder flown was ALSE (Apollo Lunar Sounder Experiment) on board Apollo 17 in 1972.

Other sounder instruments flown (in this case around Mars), are MARSIS (Mars Advanced Radar for SubSurface and Ionosphere Sounding) on board the European Space Agency's Mars Express probe, and SHARAD (mars SHAllow RADar sounder) on JPL's Mars Reconnaissance Orbiter (MRO). Both are currently operational. A radar sounder is also used on the Japanese moon probe SELENE, launched September 14, 2007.

A similar instrument (primarily devoted to ionospheric plasma probing) was embarked on the Japanese martian mission Nozomi (launched in 1998 but lost).

Secondary Surveillance Radar

SSR antenna of Deutsche Flugsicherung at Neubrandenburg, in Mecklenburg/Western Pomerania

Secondary surveillance radar (SSR) is a radar system used in air traffic control (ATC), that not only detects and measures the position of aircraft i.e. bearing, but also requests additional information from the aircraft itself such as its identity and altitude. Unlike primary radar systems that measure the bearing of targets using the detected reflections of radio signals, SSR relies on targets equipped with a radar transponder, that replies to each interrogation signal by transmitting a response containing encoded data. SSR is based on the military identification friend or foe (IFF) technology originally developed during World War II, therefore the two systems are still compatible. Monopulse secondary surveillance radar (MSSR), Mode S, TCAS and ADS-B are similar modern methods of secondary surveillance.

Overview

Primary Radar

The rapid wartime development of radar had obvious applications for air traffic control (ATC) as a means of providing continuous surveillance of air traffic disposition. Precise knowledge of the positions of aircraft would permit a reduction in the normal procedural separation standards, which in turn promised considerable increases in the efficiency of the airways system. This type of radar (now called a *primary radar*) can detect and report the position of anything that reflects its transmitted radio signals including, depending on its design, aircraft, birds, weather and land features. For air traffic control purposes this is both an advantage and a disadvantage. Its targets do not have to co-operate, they only have to be within its coverage and be able to reflect radio waves, but it only indicates the position of the targets, it does not identify them. When primary radar was the only type of radar available, the correlation of individual radar returns with specific aircraft typically was achieved by the controller observing a directed turn by the aircraft. Primary radar is still used by ATC today as a backup/complementary system to secondary radar, although its coverage and information is more limited.

Secondary Radar

Installation of mode S antenna on top of existing primary antenna

The need to be able to identify aircraft more easily and reliably led to another wartime radar development, the Identification Friend or Foe (IFF) system, which had been created as a means of positively identifying friendly aircraft from enemy. This system, which became known in civil use as secondary surveillance radar (SSR), or in the USA as the air traffic control radar beacon system (ATCRBS), relies on a piece of equipment aboard the aircraft known as a "transponder." The transponder is a radio receiver and transmitter pair which receives on 1030 MHz and transmits on 1090 MHz. The target aircraft transponder replies to signals from an interrogator (usually, but not necessarily, a ground station co-located with a primary radar) by transmitting a coded reply signal containing the requested information.

Independent secondary surveillance radar (ISSR), designation YMT, north of Chibougamau, Quebec, Canada

Both the civilian SSR and the military IFF have become much more complex than their war-time ancestors, but remain compatible with each other, not least to allow military aircraft to operate in civil airspace. Today's SSR can provide much more detailed information, for example, the aircraft altitude, as well as enabling the direct exchange of data between aircraft for collision avoidance. Most SSR systems rely on Mode C transponders, which report the aircraft pressure altitude. The pressure altitude is independent from the pilot's altimeter setting, thus preventing false altitude transmissions if altimeter is adjusted incorrectly. Air traffic control systems recalculate reported pressure altitudes to true altitudes based on their own pressure references, if necessary.

Given its primary military role of reliably identifying friends, IFF has much more secure (encrypted) messages to prevent "spoofing" by the enemy, and is used on many types of military platforms including air, sea and land vehicles.

Standards and Specifications

The International Civil Aviation Organization (ICAO) is a branch of the United Nations and its headquarters are in Montreal, Canada. It publishes annexes to the Convention and Annex 10 addresses Standards and Recommended Practices for Aeronautical Telecommunications. The objective is to ensure that aircraft crossing international boundaries are compatible with the Air Traffic Control systems in all countries that may be visited. Volume III, Part 1 is concerned with digital data communication systems including the data link functions of Mode S while volume IV defines its operation and signals in space.

The American Radio Technical Commission for Aeronautics (RTCA) and the European Organization for Civil Aviation Equipment (Eurocae) produce *Minimum Operational Performance Standards* for both ground and airborne equipment in accordance with the standards specified in ICAO Annex 10. Both organisations frequently work together and produce common documents.

ARINC (Aeronautical Radio, Incorporated) is an airline run organisation concerned with the form, fit and function of equipment carried in aircraft. Its main purpose is to ensure competition between manufacturers by specifying the size, power requirements, interfaces and performance of equipment to be located in the equipment bay of the aircraft.

Operation

The purpose of SSR is to improve the ability to detect and identify aircraft while automatically providing the Flight Level (pressure altitude) of an aircraft. An SSR ground station transmits interrogation pulses on 1030 MHz (continuously in Modes A, C and selectively, in Mode S) as its antenna rotates, or is electronically scanned, in space. An aircraft transponder within line-of-sight range 'listens' for the SSR interrogation signal and transmits a reply on 1090 MHz that provides aircraft information. The reply sent depends on the interrogation mode. The aircraft is displayed as a tagged icon on the controller's radar screen at the measured bearing and range. An aircraft without an operating transponder still may be observed by primary radar, but would be displayed to the controller without the benefit of SSR derived data. It is typically a requirement to have a working transponder in order to fly in controlled air space and many aircraft have a back-up transponder to ensure that condition is met.

Interrogation Modes

There are several modes of interrogation, each indicated by the difference in spacing between two transmitter pulses, known as P1 and P3. Each mode produces a different response from the aircraft. A third pulse, P2, is for side lobe suppression and is described later. Not included are additional military (or IFF) modes, which are described in Identification Friend or Foe.

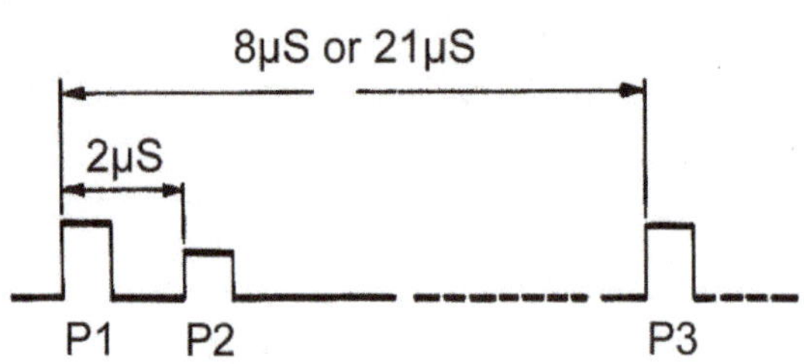

Mode A and C interrogation format

Mode	P1–P3 Pulse spacing	Purpose
A	8 μs	identity
B	17 μs	identity
C	21 μs	altitude
D	25 μs	undefined
S	3.5 μs	multipurpose

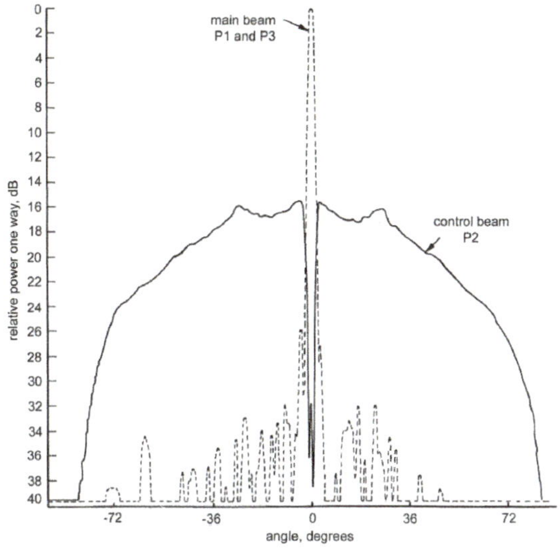

Sum and Control antenna beams

A mode-A interrogation elicits a 12-pulse reply, indicating an identity number associated with that aircraft. The 12 data pulses are bracketed by two framing pulses, F1 and F2. The X pulse is not used. A mode-C interrogation produces an 11-pulse response (pulse D1 is not used), indicating aircraft altitude as indicated by its altimeter in 100-foot increments. Mode B gave a similar response to mode A and was at one time used in Australia. Mode D has never been used operationally.

The new mode, Mode S, has different interrogation characteristics. It comprises pulses P1 and P2 from the antenna main beam to ensure that Mode-A and Mode-C transponders do not reply, followed by a long phase-modulated pulse.

The ground antenna is highly directional but cannot be designed without sidelobes. Aircraft could also detect interrogations coming from these sidelobes and reply appropriately. However these replies can not be differentiated from the intended replies from the main beam and can give rise to a false aircraft indication at an erroneous bearing. To overcome this problem the ground antenna is provided with a second, mainly omni-directional, beam with a gain which exceeds that of the sidelobes but not that of the main beam. A third pulse, P2, is transmitted from this second beam 2 μs after P1. An aircraft detecting P2 stronger than P1 (therefore in the sidelobe and at the incorrect main lobe bearing), does not reply.

Deficiencies

A number of problems are described in an ICAO publication of 1983 entitled *Secondary Surveillance Radar Mode S Advisory Circular*.

Mode A

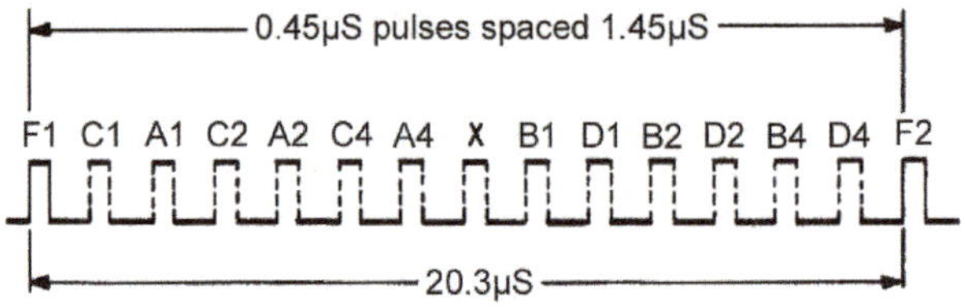

Mode A and C reply format

Although 4,096 different identity codes available in a mode A reply may seem enough, once particular codes have been reserved for emergency and other purposes, the number is significantly reduced. Ideally an aircraft would keep the same code from take-off until landing even when crossing international boundaries, as it is used at the air traffic control centre to display the aircraft's callsign using a process known as code/callsign conversion. Clearly the same mode A code should not be given to two aircraft at the same time as the controller on the ground could be given the wrong callsign with which to communicate with the aircraft.

Mode C

The mode C reply provides height increments of 100 feet, which was initially adequate for monitoring aircraft separated by at least 1000 feet. However, as airspace became increasingly congested, it became important to monitor whether aircraft were not moving out of their assigned flight level. A slight change of a few feet could cross a threshold and be indicated as the next increment up and a change of 100 feet. Smaller increments were desirable.

FRUIT

Since all aircraft reply on the same frequency of 1090 MHz, a ground station will also receive aircraft replies originating from responses to other ground stations. These unwanted replies are known as FRUIT (False Replies Unsynchronized with Interrogator Transmissions or alternatively False Replies Unsynchronized In Time). Several successive fruit replies could combine and appear to indicate an aircraft which does not exist. As air transport expands and more aircraft occupy the airspace, the amount of FRUIT generated will also increase.

Garble

FRUIT replies can overlap with wanted replies at a ground receiver, thus causing errors in extracting the included data. A solution is to increase the interrogation rate so as to receive more replies, in the hope that some would be clear of interference. The process is self-defeating as increasing the reply rate only increases the interference to other users and vice versa.

Synchronous Garble

If two aircraft paths cross within about two miles slant range from the ground interrogator, their replies will overlap and the interference caused will make their detection difficult. Typically the controller will lose the longer range, and later to reply, aircraft just when the former may be most interested in monitoring them closely.

Capture

While an aircraft is replying to one ground interrogation it is unable to respond to another interrogation, reducing detection efficiency. For a Mode A or C interrogation the transponder reply may take up to 120 μs before it can reply to a further interrogation.

Antenna

Original SSR antenna providing a narrow horizontal beam and a wide vertical beam

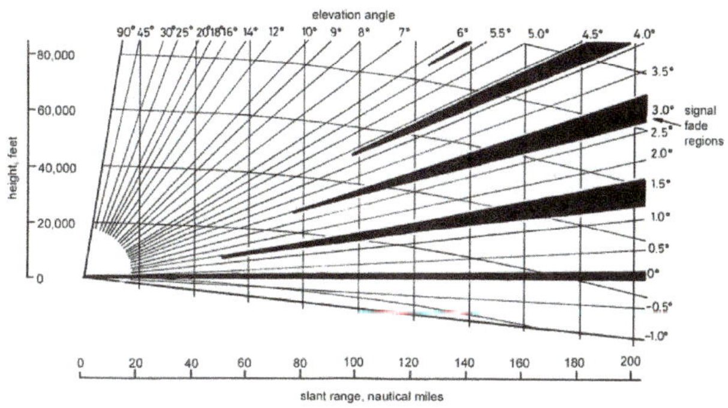

Regions of weak signal due to ground reflection

The ground antenna has a typical horizontal 3 dB beamwidth of 2.5° which limits the accuracy in determining the bearing of the aircraft. Accuracy can be improved by making many interrogations as the antenna beam scans an aircraft and a better estimate can be obtained by noting where the replies started and where stopped and taking the centre of the replies as the direction of the aircraft. This is known as a sliding window process.

The early system used an antenna known as a *hogtrough*. This has a large horizontal dimension to produce a narrow horizontal beam and a small vertical dimension to provide cover from close to the horizon to nearly overhead. There were two problems with this antenna. First, nearly half the energy is directed at the ground where it is reflected back up, and interferes with, the upward energy causing deep nulls at certain elevation angles and loss of contact with aircraft. Second, if the surrounding ground is sloping, then the reflected energy is partly offset horizontally, distorting the beam shape and the indicated bearing of the aircraft. This was particularly important in a monopulse system with its much improved bearing measurement accuracy.

Developments to Address the Deficiencies

The deficiencies in modes A and C were recognised quite early in the use of SSR and in 1967 Ullyatt published a paper and in 1969 an expanded paper, which proposed improvements to SSR to address the problems. The essence of the proposals was new interrogation and reply formats. Aircraft identity and altitude were to be included in the one reply so collation of the two data items would not be needed. Monopulse would be used to determine the bearing of the aircraft thereby reducing to one the number of interrogations/replies per aircraft on each scan of the antenna. Further each interrogation would be preceded by main beam pulses P1 and P2 separated by 2 μs so that transponders operating on modes A and C would take it as coming from the antenna sidelobe and not reply and not cause unnecessary fruit.

The FAA were also considering similar problems but were assuming that a new pair of frequencies would be required. Ullyatt showed that the existing 1030 MHz and 1090 MHz frequencies could be retained and the existing ground interrogators and airbornes transponders, with suitable modifications, could be used. The result was a Memorandum of Understanding between the US and the UK to develop a common system. In the US the programme was called DABS (Discrete Address Beacon System), and in the UK Adsel (Address selective).

Monopulse, which means single pulse, had been used in military track-and-follow systems whereby the antenna was steered to follow a particular target by keeping the target in the centre of the beam. Ullyatt proposed the use of a continuously rotating beam with bearing measurement made wherever the pulse may arrive in the beam.

The FAA engaged Lincoln Laboratory of MIT to further design the system and it produced a series of ATC Reports defining all aspects of the new joint development. Notable additions to the concept proposed by Ullyatt was the use of a more powerful 24-bit parity system using a cyclic redundancy code, which not only ensured the accuracy of the received data without the need for repetition but also allowed errors caused by an overlapping fruit reply to be corrected. Further the proposed aircraft identity code also comprised 24 bits with 16 million permutations. This allowed each aircraft to be wired with its own unique address. Blocks of addresses are allocated to different countries and further allocated to particular airlines so that knowledge of the address could identify a particular aircraft. The Lincoln Laboratory report ATC 42 entitled *Mode S Beacon System: Functional Description* gave details on the proposed new system.

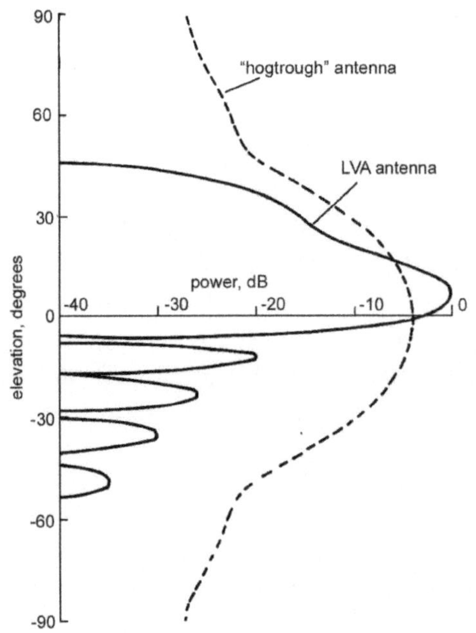

Comparison of the vertical beam shapes of the old and new antennas

The two countries reported the results of their development in a joint paper, *ADSEL/DABS – A Selective Address Secondary Surveillance Radar*. This was followed at a conference at ICAO Headquarters in Montreal, at which a low-power interrogation constructed by Lincoln Laboratory successfully communicated with an upgraded commercial SSR transponder of UK manufacture.

The only thing needed was an international name. Much had been made of the proposed new features but the existing ground SSR interrogators would still be used, albeit with modification, and the existing airbound transponders, again with modification. The best way of showing that this was an evolution not a revolution was to still call it SSR but with a new mode letter. Mode S was the obvious choice, with the S standing for select. In 1983 ICAO issued an advisory circular, which described the new system.

Improved Antenna

The problem with the existing standard "hogtrough" antenna was caused by the energy radiated toward the ground, which was reflected up and interfered with the upwards directed energy. The answer was to shape the vertical beam. This necessitated a vertical array of dipoles suitably fed to produce the desired shape. A five-foot vertical dimension was found to be optimum and this has become the international standard.

Monopulse Secondary Surveillance Radar

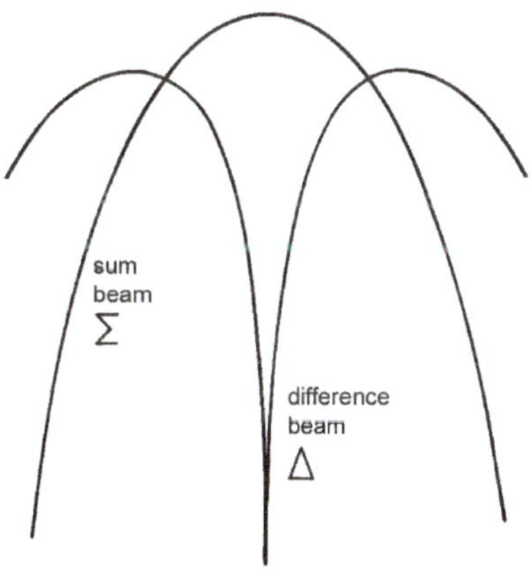

Antenna main beam with difference beam

The new Mode S system was intended to operate with just a single reply from an aircraft, a system known as monopulse. The accompanying diagram shows a conventional main or "sum" beam of an SSR antenna to which has been added a "difference" beam. To produce the sum beam the signal is distributed horizontally across the antenna aperture. This feed system is divided into two equal halves and the two parts summed again to produce the original sum beam. However the two halves are also subtracted to produce a difference output. A signal arriving exactly normal, or boresight, to the antenna will produce a maximum output in the sum beam but a zero signal in the difference beam. Away from boresight the signal in the sum beam will be less but there will be a non-zero signal in the difference beam. The angle of arrival of the signal can be determined by measuring the ratio of the signals between the sum and difference beams. The ambiguity about boresight can be resolved as there is a 180° phase change in the difference signal either side of boresight. Bearing measurements can be made on a single pulse, hence monopulse, but accuracy can be improved by averaging measurements made on several or all of the pulses received in a reply from an aircraft. A monopulse receiver was developed early in the UK Adsel programme and this design is still used widely today. Mode S reply pulses are deliberately designed to be similar to mode A and C replies so the same receiver can be used to provide improved bearing measurement for the SSR mode A and C system with the advantage that the interrogation rate can be substantially reduced thereby reducing the interference caused to other users of the system.

Lincoln Laboratory exploited the availability of a separate bearing measurement on each reply pulse to overcome some of the problems of garble whereby two replies overlap making associating the pulses with the two replies. Since each pulse is separately labelled with direction this information can be used to unscramble two overlapping mode A or C replies. The process is presented in ATC-65 "The ATCRBS Mode of DABS". The approach can be taken further by also measuring the strength of each reply pulse and using that as a discriminate as well. The following table compares the performance of conventional SSR, monopulse SSR (MSSR) and Mode S.

	Standard SSR	**Monopulse SSR**	**Mode S**
Replies per scan	20–30	4–8	1
Range accuracy	230 m rms	13 m rms	7 m rms
Bearing accuracy	0.08° rms	0.04° rms	0.04° rms
Height resolution	100 ft (30 m)	100 ft	25 ft (7.6 m)
Garble resistance	poor	good	best
Data capacity (uplink)	0	0	56–1,280 bits
Data capacity (downlink)	23 bits	23 bits	56–1,280 bits
Identity permutations	4,096	4,096	16 million

The MSSR replaced most of the existing SSRs by the 1990s and its accuracy provided for a reduction of separation minima in en-route ATC from 10 nautical miles (19 km; 12 mi) to 5 nautical miles (9.3 km; 5.8 mi).

MSSR resolved many of the system problems of SSR, as changes to the ground system only, were required. The existing transponders installed in aircraft were unaffected. It undoubtedly resulted in the delay of Mode S.

Mode S

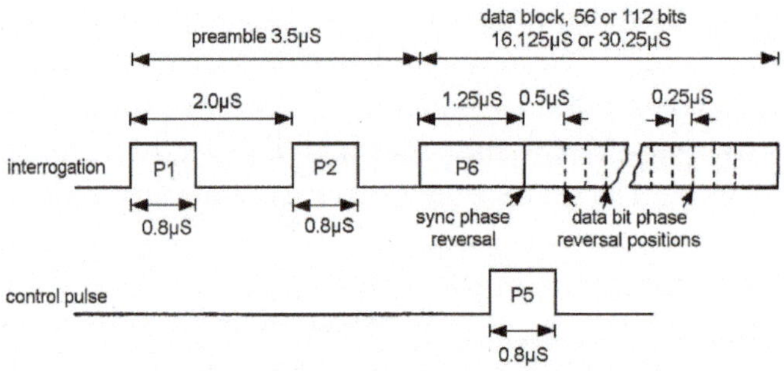

Mode S interrogation, short and long

A more detailed description of Mode S is given in the Eurocontrol publication *Principles of Mode S and Interrogator Codes* and the ICAO circular 174-AN/110 *Secondary Surveillance Radar Mode*

S Advisory Circular. The 16 million permutations of the 24 bit aircraft address codes have been allocated in blocks to individual states and the assignment is given in ICAO Annex 10, Volume III, Chapter 9.

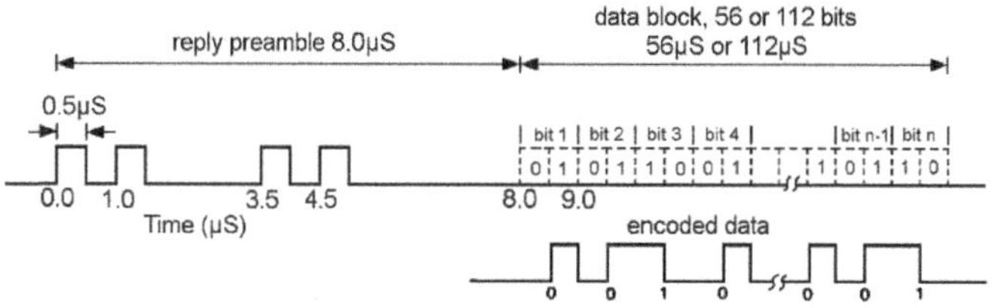

Mode S reply, short and long

A mode S interrogation comprises two 0.8 µs wide pulses, which are interpreted by a mode A & C transponder as coming from an antenna sidelobe and therefore a reply is not required. The following long P6 pulse is phase modulated with the first phase reversal, after 1.25 µs, synchronising the transponder's phase detector. Subsequent phase reversals indicate a data bit of 1, with no phase reversal indicating a bit of value 0. This form of modulation provides some resistance to corruption by a chance overlapping pulse from another ground interrogator. The interrogation may be short with P6 = 16.125 µs, mainly used to obtain a position update, or long, P6 = 30.25 µs, if an additional 56 data bits are included. The final 24 bits contain both the parity and address of the aircraft. On receiving an interrogation, an aircraft will decode the data and calculate the parity. If the remainder is not the address of the aircraft then either the interrogation was not intended for it or it was corrupted. In either case it will not reply. If the ground station was expecting a reply and did not receive one then it will re-interrogate.

The aircraft reply consists of a preamble of four pulses spaced so that they cannot be erroneously formed from overlapping mode A or C replies. The remaining pulses contain data using pulse position amplitude modulation. Each 1 µs interval is divided into two parts. If a 0.5 µs pulse occupies the first half and there is no pulse in the second half then a binary 1 is indicated. If it is the other way round then it represents a binary 0. In effect the data is transmitted twice, the second time in inverted form. This format is very resistant to error due to a garbling reply from another aircraft. To cause a hard error one pulse has to be cancelled and a second pulse inserted in the other half of the bit period. Much more likely is that both halves are confused and the decoded bit is flagged as "low confidence".

The reply also has parity and address in the final 24 bits. The ground station tracks the aircraft and uses the predicted position to indicate the range and bearing of the aircraft so it can interrogate again and get an update of its position. If it is expecting a reply and if it receives one then it checks the remainder from the parity check against the address of the expected aircraft. If it is not the same then either it is the wrong aircraft and a re-interrogation is necessary, or the reply has been corrupted by interference by being garbled by another reply. The parity system has the power to correct errors as long as they do not exceed 24 µs, which embraces the duration of a mode A or C reply, the most expected source of interference in the early days of Mode S. The pulses in the reply have individual monopulse angle measurements available, and in some implementations also signal strength measurements, which can indicate bits that are inconsistent with the majority of the other bits, thereby indicating possible corruption. A test is made by inverting the state of some or all of these bits (a 0 changed to a 1 or vice versa) and if the parity check now succeeds the changes are made permanent and the reply accepted. If it fails then a re-interrogation is required.

Mode S operates on the principle that interrogations are directed to a specific aircraft using that aircraft's unique address. This results in a single reply with aircraft range determined by the time taken to receive the reply and monopulse providing an accurate bearing measurement. In order to interrogate an aircraft its address must be known. To meet this requirement the ground interrogator also broadcasts All-Call interrogations, which are in two forms.

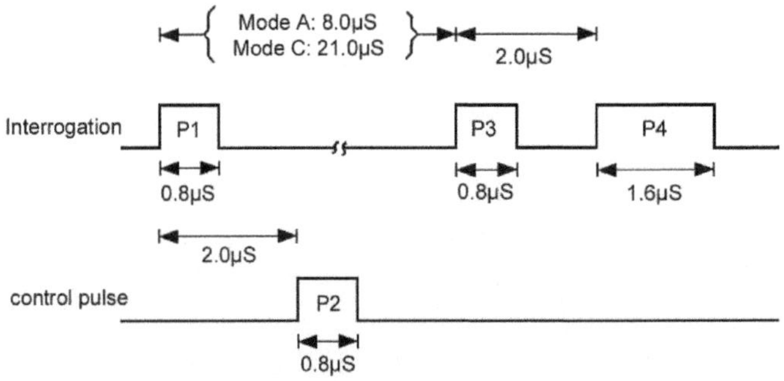

Mode A/C/S All-Call interrogation

In one form, the Mode A/C/S All-Call looks like a conventional Mode A or C interrogation at first and a transponder will start the reply process on receipt of pulse P3. However a Mode S transponder will abort this procedure upon the detection of pulse P4, and instead respond with a short Mode S reply containing its 24 bit address. This form of All-Call interrogation is now not much used as it will continue to obtain replies from aircraft already known and give rise to unnecessary interference. The alternative form of All-Call uses short Mode S interrogation with a 16.125 µs data block. This can include an indication of the interrogator transmitting the All-Call with the request that if the aircraft has already replied to this interrogator then do not reply again as aircraft is already known and a reply unnecessary.

The Mode S interrogation can take three forms:

name	form	use
Surveillance	short	position update
Comm-A	long	contains 56 data bits
Comm-C	long	up to 16 long interrogations strung together to transmit up to 1280 bits

The first five bits, known as the uplink field (UF) in the data block indicate the type of interrogation. The final 24 bits in each case is combined aircraft address and parity. Not all permutations have yet been allocated but those that have are shown:

UF	application
00000	short air-air surveillance (TCAS)
00100	surveillance, altitude request
00101	surveillance, Mode A identity request}
01011	Mode S only All-Call
10000	long air-air surveillance (TCAS)
10100	Comm-A including altitude request
10101	Comm-A including Mode A identity request
11	Comm-C (extended length message)

Similarly the Mode S reply can take three forms:

name	form	use
Surveillance	short	position update
Comm-B	long	contains 56 data bits
Comm-D	long	up to 16 long interrogations strung together to transmit up to 1280 bits

The first five bits, known as the downlink field (DF) in the data block indicate the type of reply. The final 24 bits in each case is combined aircraft address and parity. Eleven permutations have been allocated.

DF binary	DF decimal	application
00000	0	short air-air surveillance (TCAS)
00100	4	surveillance, altitude reply
00101	5	surveillance, Mode A identity reply
01011	11	All-Call reply containing aircraft address
10000	16	long air-air surveillance (TCAS)
10001	17	extended squitter
10010	18	TIS-B
10011	19	military extended squitter
10100	20	Comm-B including altitude reply
10101	21	Comm-B reply including Mode A identity
10110	22	military use
11	24	up to 16 long replies strung together to transmit up to 1280 bits

A transponder equipped to transmit Comm-B replies is fitted with 256 data registers each of 56 bits. The contents of these registers are filled and maintained from on-board data sources. If the ground system requires this data then it requests it by a Surveillance or Comm-A interrogation.

ICAO Annex 10 Volume III, Chapter 5 lists the contents of all those currently allocated. A reduced number are required for current operational use. Other registers are intended for use with TCAS and ADS-B. The Comm-B Data Selector (BDS) numbers are in hexadecimal notation.

register	data
BDS 6,0	magnetic heading
BDS 6,0	indicated airspeed
BDS 6,0	Mach number
BDS 6,0	vertical rate
BDS 5,0	roll angle
BDS 5,0	track angle rate
BDS 5,0	true track angle
BDS 5,0	ground speed
BDS 4,0	selected vertical intent

Extended Squitter

Starting in 2009, the ICAO defined an "extended squitter" mode of operation; it supplements the requirements contained in ICAO Annex 10, Volumes III and IV. The first edition specified earlier versions of extended squitter messages:

Version 0

Extends Mode S to deal with basic ADS-B exchanges, to add traffic information broadcast (TIS-B) format information, as well as uplink and downlink broadcast protocol information.

Version 1

Better describes surveillance accuracy and integrity information (navigation accuracy category, navigation integrity category, surveillance integrity level), and additional parameters for TIS-B and ADS-B rebroadcast (ADS-R).

Version 2

The second edition introduced yet a new version of extended squitter formats and protocols to:

- Enhance integrity and accuracy reporting.
- Add a number of additional parameters to support identified operational needs for the use of ADS-B not covered by Version 1 (including capabilities to support airport surface applications).
- Modify several parameters, and remove a number of parameters, which are no longer required to support ADS-B applications.

Radar Tracker

A radar tracker is a component of a radar system, or an associated command and control (C2) system, that associates consecutive radar observations of the same target into tracks. It is particularly useful when the radar system is reporting data from several different targets or when it is necessary to combine the data from several different radars or other sensors.

Role of the Radar Tracker

A classical rotating air surveillance radar system detects target echoes against a background of noise. It reports these detections (known as "plots") in polar coordinates representing the range and bearing of the target. In addition, noise in the radar receiver will occasionally exceed the detection threshold of the radar's Constant false alarm rate detector and be incorrectly reported as targets (known as false alarms). The role of the radar tracker is to monitor consecutive updates from the radar system (which typically occur once every few seconds, as the antenna rotates) and to determine those sequences of plots belonging to the same target, whilst rejecting any plots believed to be false alarms. In addition, the radar tracker is able to use the sequence of plots to estimate the current speed and heading of the target. When several targets are present, the radar tracker aims to provide one track for each target, with the track history often being used to indicate where the target has come from.

When multiple radar systems are connected to a single reporting post, a multiradar tracker is often used to monitor the updates from all of the radars and form tracks from the combination of detections. In this configuration, the tracks are often more accurate than those formed from single radars, as a greater number of detections can be used to estimate the tracks. In addition to associating plots, rejecting false alarms and estimating heading and speed, the radar tracker also acts as a filter, in which errors in the individual radar measurements are smoothed out. In essence, the radar tracker fits a smooth curve to the reported plots and, if done correctly, can increase the overall accuracy of the radar system. A multisensor tracker extends the concept of the multiradar tracker to allow the combination of reports from different types of sensor - typically radars, secondary surveillance radars (SSR), identification friend or foe (IFF) systems and electronic support measures (ESM) data.

A radar track will typically contain the following information:

- Position (in two or three dimensions)
- Heading
- Speed
- Unique track number

In addition, and depending on the application or tracker sophistication, the track will also include:

- Civilian SSR Modes A, C, S information
- Military IFF Modes 1, 2, 3, 4 and 5 information
- Call sign information
- Track reliability or uncertainty information

General Approach

There are many different mathematical algorithms used for implementing a radar tracker, of varying levels of sophistication. However, they all perform steps similar to the following every time the radar updates:

- Associate a radar plot with an existing track (plot to track association)
- Update the track with this latest plot (track smoothing)
- Spawn new tracks with any plots that are not associated with existing tracks (track initiation)
- Delete any tracks that have not been updated, or predict their new location based on the previous heading and speed (track maintenance)

Perhaps the most important step is the updating of tracks with new plots. All trackers will implicitly or explicitly take account of a number of factors during this stage, including:

- A model for how the radar measurements are related to the target coordinates
- The errors on the radar measurements
- A model of the target movement
- Errors in the model of the target movement

Using these information, the radar tracker attempts to update the track by forming a weighted average of the current reported position from the radar (which has unknown errors) and the last predicted position of the target from the tracker (which also has unknown errors). The tracking problem is made particularly difficult for targets with unpredictable movements (i.e. unknown target movement models), non-Gaussian measurement or model errors, non-linear relationships between the measured quantities and the desired target coordinates, detection in the presence of non-uniformly distributed clutter, missed detections or false alarms. In the real world, a radar tracker typically faces a combination of all of these effects; this has led to the development of an increasingly sophisticated set of algorithms to resolve the problem. Due to the need to form radar tracks in real time, usually for several hundred targets at once, the deployment of radar tracking algorithms has typically been limited by the available computational power.

Plot to Track Association

In this step of the processing, the radar tracker seeks to determine which plots should be used to update which tracks. In many approaches, a given plot can only be used to update one track. However, in other approaches a plot can be used to update several tracks, recognising the uncertainty in knowing to which track the plot belongs. Either way, the first step in the process is to update all of the existing tracks to the current time by predicting their new position based on the most recent state estimate (e.g. position, heading, speed, acceleration, etc.) and the assumed target motion model (e.g. constant velocity, constant acceleration, etc.). Having updated the estimates, it is possible to try to associate the plots to tracks.

This can be done in a number of ways:

- By defining an "acceptance gate" around the current track location and then selecting:
  - the closest plot in the gate to the predicted position, or
  - the strongest plot in the gate
- By a statistical approach, such as the Probabilistic Data Association Filter (PDAF) or the Joint Probabilistic Data Association Filter (JPDAF) that choose the most probable location of plot through a statistical combination of all the likely plots. This approach has been shown to be good in situations of high radar clutter.

Once a track has been associated with a plot, it moves to the track smoothing stage, where the track prediction and associated plot are combined to provide a new, smoothed estimate of the target location.

Having completed this process, a number of plots will remain unassociated to existing tracks and a number of tracks will remain without updates. This leads to the steps of track initiation and track maintenance.

Track Initiation

Track initiation is the process of creating a new radar track from an unassociated radar plot. When the tracker is first switched on, all the initial radar plots are used to create new tracks, but once the tracker is running, only those plots that couldn't be used to update an existing track are used to spawn new tracks. Typically a new track is given the status of tentative until plots from subsequent radar updates have been successfully associated with the new track. Tentative tracks are not shown to the operator and so they provide a means of preventing false tracks from appearing on the screen - at the expense of some delay in the first reporting of a track. Once several updates have been received, the track is confirmed and displayed to the operator. The most common criterion for promoting a tentative track to a confirmed track is the "M-of-N rule", which states that during the last N radar updates, at least M plots must have been associated with the tentative track - with M=3 and N=5 being typical values. More sophisticated approaches may use a statistical approach in which a track becomes confirmed when, for instance, its covariance matrix falls to a given size.

Track Maintenance

Track maintenance is the process in which a decision is made about whether to end the life of a track. If a track was not associated with a plot during the plot to track association phase, then there is a chance that the target may no longer exist (for instance, an aircraft may have landed or flown out of radar cover). Alternatively, however, there is a chance that the radar may have just failed to see the target at that update, but will find it again on the next update. Common approaches to deciding on whether to terminate a track include:

- If the target was not seen for the past M consecutive update opportunities (typically M=3 or so).
- If the target was not seen for the past M out of N most recent update opportunities.
- If the target's track uncertainty (covariance matrix) has grown beyond a certain threshold.

Track Smoothing

In this important step, the latest track prediction is combined with the associated plot to provide a new, improved estimate of the target state as well as a revised estimate of the errors in this prediction. There is a wide variety of algorithms, of differing complexity and computational load, that can be used for this process.

Alpha-beta Tracker

An early tracking approach, using an alpha beta filter, that assumed fixed covariance errors and a constant-speed, non-maneuvering target model to update tracks.

Kalman Filter

The role of the Kalman Filter is to take the current known state (i.e. position, heading, speed and possibly acceleration) of the target and predict the new state of the target at the time of the most recent radar measurement. In making this prediction, it also updates its estimate of its own uncertainty (i.e. errors) in this prediction. It then forms a weighted average of this prediction of state and the latest measurement of state, taking account of the known measurement errors of the radar and its own uncertainty in the target motion models. Finally, it updates its estimate of its uncertainty of the state estimate. A key assumption in the mathematics of the Kalman filter is that measurement equations (i.e. the relationship between the radar measurements and the target state) and the state equations (i.e. the equations for predicting a future state based on the current state) are linear.

The Kalman filter assumes that the measurement errors of the radar, and the errors in its target motion model, and the errors in its state estimate are all zero-mean with known covariance. This means that all of these sources of errors can be represented by a covariance matrix. The mathematics of the Kalman filter is therefore concerned with propagating these covariance matrices and using them to form the weighted sum of prediction and measurement.

In situations where the target motion conforms well to the underlying model, there is a tendency of the Kalman filter to become "overconfident" of its own predictions and to start to ignore the radar measurements. If the target then manoeuvres, the filter will fail to follow the manoeuvre. It is therefore common practice when implementing the filter to arbitrarily increase the magnitude of the state estimate covariance matrix slightly at each update to prevent this.

Multiple Hypothesis Tracker (MHT)

The MHT allows a track to be updated by more than one plot at each update, spawning multiple possible tracks. As each radar update is received every possible track can be potentially updated with every new update. Over time, the track branches into many possible directions. The MHT calculates the probability of each potential track and typically only reports the most probable of all the tracks. For reasons of finite computer memory and computational power, the MHT typically includes some approach for deleting the most unlikely potential track updates. The MHT is designed for situations in which the target motion model is very unpredictable, as all potential track updates are considered. For this reason, it is popular for problems of ground target tracking in Airborne Ground Surveillance (AGS) systems.

Interacting Multiple Model (IMM)

The IMM is an estimator which can either be used by MHT or JPDAF. IMM uses two or more Kalman filters which run in parallel, each using a different model for target motion or errors. The IMM forms an optimal weighted sum of the output of all the filters and is able to rapidly adjust to target maneuvers. While MHT or JPDAF handles the association and track maintenance, an IMM helps MHT or JPDAF in obtaining a filtered estimate of the target position.

Nonlinear Tracking Algorithms

Non-linear tracking algorithms use a Non-linear filter to cope with the situation where the measurements have a non-linear relationship to the final track coordinates, where the errors are non-Gaussian, or where the motion update model is non-linear. The most common non-linear filters are:

- the Extended Kalman filter
- the Unscented Kalman filter
- the Particle filter

Extended Kalman Filter (EKF)

The EKF is an extension of the Kalman filter to cope with cases where the relationship between the radar measurements and the track coordinates, or the track coordinates and the motion model, is non-linear. In this case, the relationship between the measurements and the state is of the form h = f(x) (where h is the vector of measurements, x is the target state and f(.) is the function relating the two). Similarly, the relationship between the future state and the current state is of the form x(t+1) = g(x(t)) (where x(t) is the state at time t and g(.) is the function that predicts the future state). To handle these non-linearities, the EKF linearises the two non-linear equations using the first term of the Taylor series and then treats the problem as the standard linear Kalman filter problem. Although conceptually simple, the filter can easily diverge (i.e. gradually perform more and more badly) if the state estimate about which the equations are linearised is poor.

The unscented Kalman filter and particle filters are attempts to overcome the problem of linearising the equations.

Unscented Kalman Filter (UKF)

The UKF attempts to improve on the EKF by removing the need to linearise the measurement and state equations. It avoids linearization by representing the mean and covariance information in the form of a set of points, called sigma points. These points, which represent a distribution with specified mean and covariance, are then propagated directly through the non-linear equations, and the resulting five updated samples are then used to calculate a new mean and variance. This approach then suffers none of the problems of divergence due to poor linearisation and yet retains the overall computational simplicity of the EKF.

Particle Filter

The particle filter could be considered as a generalisation of the UKF. It makes no assumptions about the distributions of the errors in the filter and neither does it require the equations to be linear. Instead it generates a large number of random potential states ("particles") and then propagates this "cloud of particles" through the equations, resulting in a different distribution of particles at the output. The resulting distribution of particles can then be used to calculate a mean or variance, or whatever other statistical measure is required. The resulting statistics are used to generate the random sample of particles for the next iteration. The particle filter is notable in its ability to handle multi-modal distributions (i.e. distributions where the PDF has more than one peak). However, it is computationally very intensive and is currently unsuitable for most real-world, real-time applications.

References

- Doviak, R. J.; Zrnic, D. S.; Sirmans, D. S. (1979). "Doppler weather radar". Proceedings of the IEEE. 67 (11): 1522–1553. doi:10.1109/PROC.1979.11511
- Shupe, M. D.; Uttal, T.; Matrosov, S. Y. (2005). "Arctic cloud microphysics retrievals from surface-based remote sensors at sheba". Journal of Applied Meteorology. 44 (10): 1544–1562. doi:10.1175/JAM2297.1
- Raju, G. S. N. (2008). Radar engineering and fundamentals of navigational aids. New Delhi: I. K. International Publishing House Pvt Ltd. pp. 54, 237, 241, 252–259. ISBN 9788190694216
- Chi M. Shun and Sharon S. Y. Lau (2000). "Terminal Doppler Weather Radar (TDWR) observation of atmospheric flow over complex terrain during tropical cyclone passages". Proc. SPIE. 4152 (42): 42. doi:10.1117/12.410622
- Martner, B. E.; Moran, K. P. (2001). "Using cloud radar polarization measurements to evaluate stratus cloud and insect echoes". Journal of Geophysical Research. 106: 4891– 4897. doi:10.1029/2000jd900623
- "Glossary of Terms". Radar - Operational Characteristics of Radar Classified by Tactical Application. pp. 109–114. Retrieved April 1, 2016
- Matrosov, S. Y. (July 1991). "Prospects for the measurement of ice cloud particle shape and orientation with elliptically polarized radar signals". Radio Science. 26 (4): 847–856. doi:10.1029/91RS00965
- de Boer, Gijs; Eloranta, Edwin W.; Shupe, Matthew D. (2009). "Arctic Mixed-Phase Stratiform Cloud Properties from Multiple Years of Surface-Based Measurements at Two High-Latitude Locations". Journal of the Atmospheric Sciences. 66 (9): 2874–2887. doi:10.1175/2009JAS3029
- Clothiaux, E .E.; Ackerman, T. P.; Babb, D. M. (1996). Ground-based remote sensing of cloud properties using millimeter-wave radar. In: Radiation and Water in the Climate System. Springer. pp. 323–366. ISBN 978-3-662-03289-3
- Pazmany, A.; McIntosh, R.; Kelly, R.; Vali, G. (1994). "An airborne 95 ghz dual-polarized radar for cloud studies". IEEE Trans. on Geosci. Rem. Sens. 32: 731–739. doi:10.1109/36.298002
- Jaffe, Bernard M., "Forward Reflection of Light by a Moving Mirror," American Journal of Physics, Vol. 41, April 1973, p577-578
- Bharadwaj, Nitin; Nelson, Dan; Lindenmaier, Iosif; Johnson, Karen. "Instrument : Ka ARM Zenith Radar (KAZR)". ARM Climate Research Facility. Retrieved 29 May 2015
- Galletti, M.; Huang, D.; Kollias, P. (2014). "Zenith/nadir pointing mm-wave radars: Linear or circular polarization?". IEEE Transactions on Geoscience and Remote Sensing. 52: 628–639. doi:10.1109/tgrs.2013.2243155

3

A Comprehensive Study of Weather Radars

Weather radar is a type of radar that is used in forecasting the weather. They are mostly pulse-Doppler radars that can detect the motion of rain droplets. The main types of radar output explained are plan position indicator, vertical composite, accumulations, echotops, range height indicator and automatic algorithms. This chapter will provide an integrated understanding of weather radars.

Weather Radar

University of Oklahoma OU-PRIME C-band, polarimetric, weather radar during construction

Weather radar, also called weather surveillance radar (WSR) and Doppler weather radar, is a type of radar used to locate precipitation, calculate its motion, and estimate its type (rain, snow, hail etc.). Modern weather radars are mostly pulse-Doppler radars, capable of detecting the motion of rain droplets in addition to the intensity of the precipitation. Both types of data can be analyzed to determine the structure of storms and their potential to cause severe weather.

During World War II, radar operators discovered that weather was causing echoes on their screen, masking potential enemy targets. Techniques were developed to filter them, but scientists began to study the phenomenon. Soon after the war, surplus radars were used to detect precipitation. Since then, weather radar has evolved on its own and is now used by national weather services, research departments in universities, and in television newscasts. Raw images are routinely used and specialized software can take radar data to make short term forecasts of future positions and intensities of rain, snow, hail, and other weather phenomena. Radar output is even incorporated into numerical weather prediction models to improve analyses and forecasts.

Weather radar in Norman, Oklahoma with rainshaft

Weather (WF44) radar dish

History

David Atlas

During World War II, military radar operators noticed noise in returned echoes due to rain, snow, and sleet. After the war, military scientists returned to civilian life or continued in the Armed Forces and pursued their work in developing a use for those echoes. In the United States, David Atlas, at first working for the Air Force and later for MIT, developed the first operational weather radars. In Canada, J.S. Marshall and R.H. Douglas formed the "Stormy Weather Group" in Montreal. Marshall and his doctoral student Walter Palmer are well known for their work on the drop size distribution in mid-latitude rain that led to understanding of the Z-R relation, which correlates a given radar reflectivity with the rate at which rainwater is falling. In the United Kingdom, research continued to study the radar echo patterns and weather elements such as stratiform rain and convective clouds, and experiments were done to evaluate the potential of different wavelengths from 1 to 10 centimeters. By 1950 the UK company EKCO was demonstrating its airborne 'cloud and collision warning search radar equipment'.

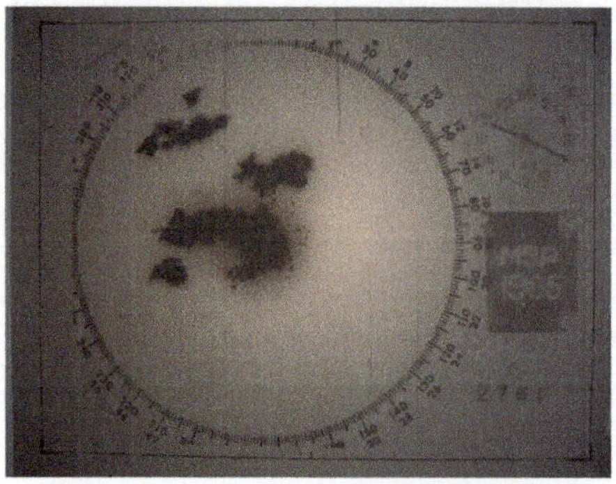

1960s radar technology detected tornado producing supercells over the Minneapolis-Saint Paul metropolitan area.

In 1953 Donald Staggs, an electrical engineer working for the Illinois State Water Survey, made the first recorded radar observation of a "hook echo" associated with a tornadic thunderstorm.

Between 1950 and 1980, reflectivity radars, which measure position and intensity of precipitation, were incorporated by weather services around the world. The early meteorologists had to watch a cathode ray tube. During the 1970s, radars began to be standardized and organized into networks. The first devices to capture radar images were developed. The number of scanned angles was increased to get a three-dimensional view of the precipitation, so that horizontal cross-sections (CAPPI) and vertical cross-sections could be performed. Studies of the organization of thunderstorms were then possible for the Alberta Hail Project in Canada and National Severe Storms Laboratory (NSSL) in the US in particular.

The NSSL, created in 1964, began experimentation on dual polarization signals and on Doppler effect uses. In May 1973, a tornado devastated Union City, Oklahoma, just west of Oklahoma City. For the first time, a Dopplerized 10 cm wavelength radar from NSSL documented the entire life cycle of the tornado. The researchers discovered a mesoscale rotation in the cloud aloft before the tornado touched the ground – the tornadic vortex signature. NSSL's research helped convince the National Weather Service that Doppler radar was a crucial forecasting tool. The Super Outbreak of tornadoes on 3–4 April 1974 and their devastating destruction might have helped to get funding for further developments.

Between 1980 and 2000, weather radar networks became the norm in North America, Europe, Japan and other developed countries. Conventional radars were replaced by Doppler radars, which in addition to position and intensity could track the relative velocity of the particles in the air. In the United States, the construction of a network consisting of 10 cm radars, called NEXRAD or WSR-88D (Weather Surveillance Radar 1988 Doppler), was started in 1988 following NSSL's research. In Canada, Environment Canada constructed the King City station, with a 5 cm research Doppler radar, by 1985; McGill University dopplerized its radar (J. S. Marshall Radar Observatory) in 1993. This led to a complete Canadian Doppler network between 1998 and 2004. France and other European countries had switched to Doppler networks by the early 2000s. Meanwhile, rapid advances in computer technology led to algorithms to detect signs of severe weather, and many applications for media outlets and researchers.

After 2000, research on dual polarization technology moved into operational use, increasing the amount of information available on precipitation type (e.g. rain vs. snow). "Dual polarization" means that microwave radiation which is polarized both horizontally and vertically (with respect to the ground) is emitted. Wide-scale deployment was done by the end of the decade or the beginning of the next in some countries such as the United States, France, and Canada. In April 2013, all National Weather Service NEXRADs were completely dual-polarized.

Since 2003, the U.S. National Oceanic and Atmospheric Administration has been experimenting with phased-array radar as a replacement for conventional parabolic antenna to provide more time resolution in atmospheric sounding. This could be significant with severe thunderstorms, as their evolution can be better evaluated with more timely data.

Also in 2003, the National Science Foundation established the Engineering Research Center for Collaborative Adaptive Sensing of the Atmosphere (CASA), a multidisciplinary, multi-university collaboration of engineers, computer scientists, meteorologists, and sociologists to conduct fundamental research, develop enabling technology, and deploy prototype engineering systems designed to augment existing radar systems by sampling the generally undersampled lower troposphere with inexpensive, fast scanning, dual polarization, mechanically scanned and phased array radars.

How a Weather Radar Works

Sending Radar Pulses

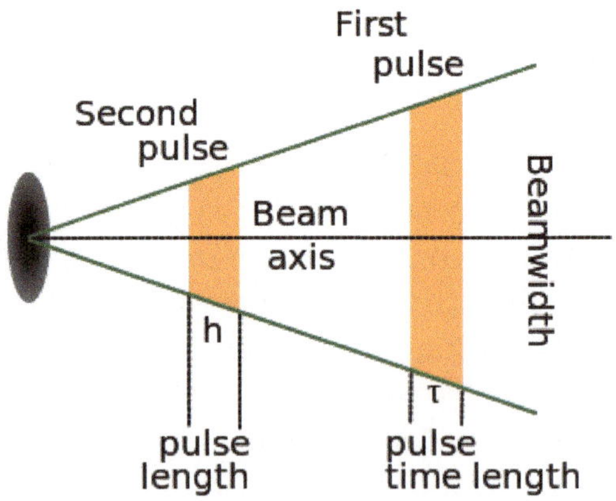

A radar beam spreads out as it moves away from the radar station, covering an increasingly large volume.

Weather radars send directional pulses of microwave radiation, on the order of a microsecond long, using a cavity magnetron or klystron tube connected by a waveguide to a parabolic antenna. The wavelengths of 1 – 10 cm are approximately ten times the diameter of the droplets or ice particles of interest, because Rayleigh scattering occurs at these frequencies. This means that part of the energy of each pulse will bounce off these small particles, back in the direction of the radar station.

Shorter wavelengths are useful for smaller particles, but the signal is more quickly attenuated. Thus 10 cm (S-band) radar is preferred but is more expensive than a 5 cm C-band system. 3 cm X-band radar is used only for short-range units, and 1 cm Ka-band weather radar is used only for research on small-particle phenomena such as drizzle and fog. W-band weather radar systems have seen limited university use, but due to quicker attenuation, most data are not operational.

Radar pulses spread out as they move away from the radar station. Thus the volume of air that a radar pulse is traversing is larger for areas farther away from the station, and smaller for nearby areas, decreasing resolution at far distances. At the end of a 150 – 200 km sounding range, the volume of air scanned by a single pulse might be on the order of a cubic kilometer. This is called the *pulse volume*

The volume of air that a given pulse takes up at any point in time may be approximated by the formula $v = hr^2\theta^2$, where v is the volume enclosed by the pulse, h is pulse width (in e.g. meters, calculated from the duration in seconds of the pulse times the speed of light), r is the distance from the radar that the pulse has already traveled (in e.g. meters), and θ is the beam width (in radians). This formula assumes the beam is symmetrically circular, “r” is much greater than “h” so “r” taken at the beginning or at the end of the pulse is almost the same, and the shape of the volume is a cone frustum of depth “h”.

Listening for Return Signals

Between each pulse, the radar station serves as a receiver as it listens for return signals from particles in the air. The duration of the "listen" cycle is on the order of a millisecond, which is a thousand times longer than the pulse duration. The length of this phase is determined by the need for the microwave radiation (which travels at the speed of light) to propagate from the detector to the weather target and back again, a distance which could be several hundred kilometers. The horizontal distance from station to target is calculated simply from the amount of time that lapses from the initiation of the pulse to the detection of the return signal. The time is converted into distance by multiplying by the speed of light in air:

$$\text{Distance} = c\frac{\Delta t}{2n},$$

where $c = 299{,}792.458$ km/s is the speed of light, and $n \approx 1.0003$ is the refractive index of air.

If pulses are emitted too frequently, the returns from one pulse will be confused with the returns from previous pulses, resulting in incorrect distance calculations.

Determining Height

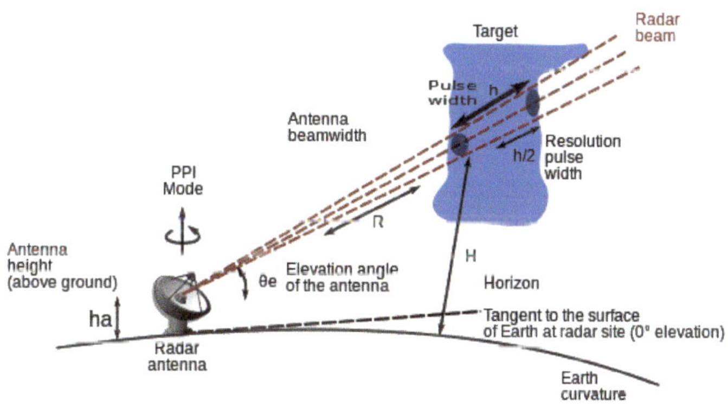

The radar beam path with height

Assuming the Earth is round, the radar beam in vacuum would rise according to the reverse curvature of the Earth. However, the atmosphere has a refractive index that diminishes with height, due to its diminishing density. This bends the radar beam slightly toward the ground and with a standard atmosphere this is equivalent to considering that the curvature of the beam is 4/3 the actual curvature of the Earth. Depending on the elevation angle of the antenna and other considerations, the following formula may be used to calculate the target's height above ground:

$$H = \sqrt{r^2 + (k_e a_e)^2 + 2rk_e a_e \sin(\theta_e)} - k_e a_e + h_a,$$

where:

r = distance radar–target,

$k_e = 4/3$,

a_e = Earth radius,

θ_e = elevation angle above the radar horizon,

h_a = height of the feedhorn above ground.

Scanned volume by using multiple elevation angles

A weather radar network uses a series of typical angles that will be set according to the needs. After each scanning rotation, the antenna elevation is changed for the next sounding. This scenario will be repeated on many angles to scan all the volume of air around the radar within the maximum range. Usually, this scanning strategy is completed within 5 to 10 minutes to have data within 15 km above ground and 250 km distance of the radar. For instance in Canada, the 5 cm weather radars use angles ranging from 0.3 to 25 degrees. The image shows the volume scanned when multiple angles are used.

Due to the Earth's curvature and change of index of refraction with height, the radar cannot "see" below the height above ground of the minimal angle (shown in green) or closer to the radar than the maximal one (shown as a red cone in the center).

Calibrating Intensity of Return

Because the targets are not unique in each volume, the radar equation has to be developed beyond the basic one. Assuming a monostatic radar where $G_t = A_r (or\ G_r) = G$:

$$P_r = P_t \frac{G^2 \lambda^2 \sigma_0}{(4\pi)^3 R^4} \propto \frac{\sigma_0}{R^4}$$

where P_r is received power, P_t is transmitted power, G is the gain of the transmitting/receiving antenna, λ is radar wavelength, σ is the radar cross section of the target and R is the distance from transmitter to target.

In this case, we have to add the cross sections of all the targets:

$$\sigma_0 = \bar{\sigma}_0 = V \sum \sigma_{0j} = V\eta$$

$$\begin{cases} V & scanned\ volume \\ & pulse\ length\ \ X\ beam\ width \\ & \frac{c\tau}{2}\frac{\pi R^2 \theta^2}{4} \end{cases}$$

where c is the light speed, τ is temporal duration of a pulse and θ is the beam width in radians.

In combining the two equations:

$$P_r = P_t \frac{G^2\lambda^2}{(4\pi)^3 R^4} \frac{c\tau}{2} \frac{\pi R^2\theta^2}{4} \eta = P_t \tau G^2 \lambda^2 \theta^2 \frac{c}{512(\pi^2)} \frac{\eta}{R^2}$$

Which leads to:

$$P_r \propto \frac{\eta}{R^2}$$

Notice that the return now varies inversely to R^2 instead of R^4. In order to compare the data coming from different distances from the radar, one has to normalize them with this ratio.

Data Types

Reflectivity (in Decibel or dBZ)

Return echoes from targets ("reflectivity") are analyzed for their intensities to establish the precipitation rate in the scanned volume. The wavelengths used (1–10 cm) ensure that this return is proportional to the rate because they are within the validity of Rayleigh scattering which states that the targets must be much smaller than the wavelength of the scanning wave (by a factor of 10).

Reflectivity perceived by the radar (Z_e) varies by the sixth power of the rain droplets' diameter (D), the square of the dielectric constant (K) of the targets and the drop size distribution (e.g. N[D] of *Marshall-Palmer*) of the drops. This gives a truncated Gamma function, of the form:

$$Z_e = \int_0^{Dmax} |K|^2 N_0 e^{-\Lambda D} D^6 dD$$

Precipitation rate (R), on the other hand, is equal to the number of particles, their volume and their fall speed (v[D]) as:

$$R = \int_0^{Dmax} N_0 e^{-\Lambda D} \frac{\pi D^3}{6} v(D) dD$$

So Z_e and R have similar functions that can be resolved giving a relation between the two of the form:

$$Z = aR^b$$

Where a and b depend on the type of precipitation (snow, rain, convective or stratiform), which has different Λ, K, N_0 and v.

- As the antenna scans the atmosphere, on every angle of azimuth it obtains a certain strength of return from each type of target encountered. Reflectivity is then averaged for that target to have a better data set.

- Since variation in diameter and dielectric constant of the targets can lead to large variability in power return to the radar, reflectivity is expressed in dBZ (10 times the logarithm of the ratio of the echo to a standard 1 mm diameter drop filling the same scanned volume).

How to Read Reflectivity on a Radar Display

Radar returns are usually described by colour or level. The colours in a radar image normally range from blue or green for weak returns, to red or magenta for very strong returns. The numbers in a verbal report increase with the severity of the returns. For example, the U.S. National Doppler Radar sites use the following scale for different levels of reflectivity:

- magenta: 65 dBZ (extremely heavy precipitation, possible hail)
- red: 52 dBZ
- yellow: 36 dBZ
- green: 20 dBZ (light precipitation)

Strong returns (red or magenta) may indicate not only heavy rain but also thunderstorms, hail, strong winds, or tornadoes, but they need to be interpreted carefully, for reasons described below.

Aviation Conventions

When describing weather radar returns, pilots, dispatchers, and air traffic controllers will typically refer to three return levels:

- level 1 corresponds to a green radar return, indicating usually light precipitation and little to no turbulence, leading to a possibility of reduced visibility.
- level 2 corresponds to a yellow radar return, indicating moderate precipitation, leading to the possibility of very low visibility, moderate turbulence and an uncomfortable ride for aircraft passengers.
- level 3 corresponds to a red radar return, indicating heavy precipitation, leading to the possibility of thunderstorms and severe turbulence and structural damage to the aircraft.

Aircraft will try to avoid level 2 returns when possible, and will always avoid level 3 unless they are specially-designed research aircraft.

Precipitation Types

Some displays provided by commercial weather sites, like *The Weather Channel*, show precipitation types during the winter month : rain, snow, mixed precipitations (sleet and freezing rain). This is not an analysis of the radar data itself but a post-treatment done with other data sources, the primary being surface reports (METAR).

Over the area covered by radar echoes, a program assigns a precipitation type according to the surface temperature and dew point reported at the underlying weather stations. Precipitation types reported by human operated stations and certain automatic ones (AWOS) will have higher weight. Then the program does interpolations to produce an image with defined zones. These will include interpolation errors due to the calculation. Mesoscale variations of the precipitation zones will also be lost. More sophisticated programs use the numerical weather prediction output from models, such as NAM and WRF, for the precipitation types and apply it as a first guess to the radar echoes, then use the surface data for final output.

Until dual-polarization data are widely available, any precipitation types on radar images are only indirect information and must be taken with care.

Velocity

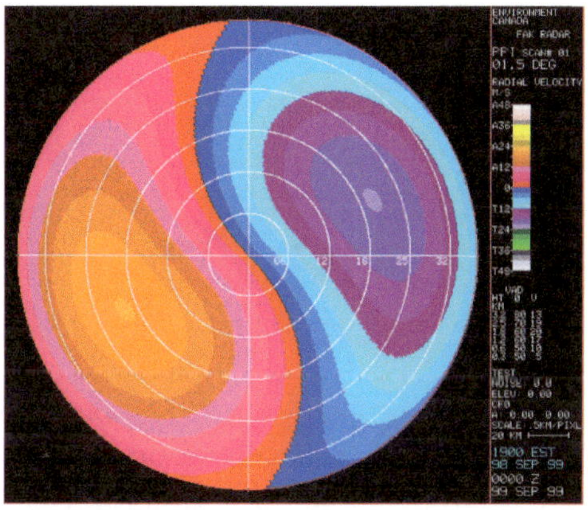

Idealized example of Doppler output. Approaching velocities are in blue and receding velocities are in red. Notice the sinusoidal variation of speed when going around the display at a particular range.

Precipitation is found in and below clouds. Light precipitation such as drops and flakes is subject to the air currents, and scanning radar can pick up the horizontal component of this motion, thus giving the possibility to estimate the wind speed and direction where precipitation is present.

A target's motion relative to the radar station causes a change in the reflected frequency of the radar pulse, due to the Doppler effect. With velocities of less than 70-metre/second for weather echos and radar wavelength of 10 cm, this amounts to a change only 0.1 ppm. This difference is too small to be noted by electronic instruments. However, as the targets move slightly between each pulse, the returned wave has a noticeable phase difference or *phase shift* from pulse to pulse.

Pulse Pair

Doppler weather radars use this phase difference (pulse pair difference) to calculate the precipitation's motion. The intensity of the successively returning pulse from the same scanned volume where targets have slightly moved is:

$$I = I_0 sin\left(\frac{4\pi(x0+v\Delta t}{\lambda}\right) = I_0 sin(\Theta_0 + \Delta\Theta) \begin{cases} x = distance\ from\ radar\ to\ target \\ \lambda = radar\ wavelength \\ \Delta t = time\ between\ two\ pulses \end{cases}$$

So $\Delta\Theta = \frac{4\pi v \Delta t}{\lambda}$, v = target speed = $\frac{\lambda \Delta\Theta}{4\pi \Delta t}$. This speed is called the radial Doppler velocity because it gives only the radial variation of distance versus time between the radar and the target. The real speed and direction of motion has to be extracted by the process described below.

Doppler Dilemma

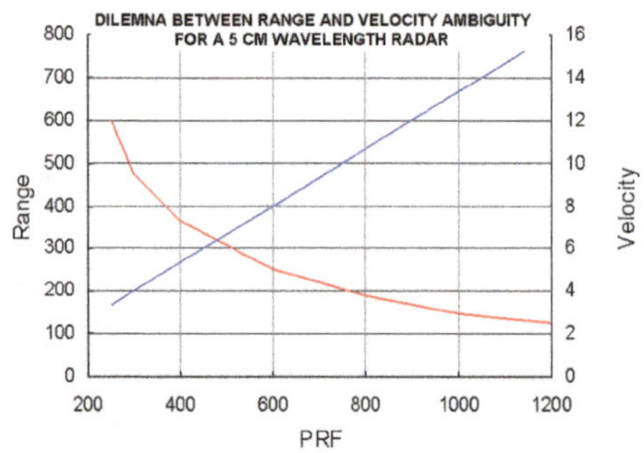

Maximum range from reflectivity (red) and unambiguous Doppler velocity range (blue) with pulse repetition frequency

The phase between pulse pairs can vary from $-\pi$ and $+\pi$, so the unambiguous Doppler velocity range is

$$V_{max} = \pm \frac{\lambda}{4\Delta t}$$

This is called the Nyquist velocity. This is inversely dependent on the time between successive pulses: the smaller the interval, the larger is the unambiguous velocity range. However, we know that the maximum range from reflectivity is directly proportional to Δt:

$$x = \frac{c\Delta t}{2}$$

The choice becomes increasing the range from reflectivity at the expense of velocity range, or increasing the latter at the expense of range from reflectivity. In general, the useful range compromise is 100–150 km for reflectivity. This means for a wavelength of 5 cm (as shown in the diagram), an unambiguous velocity range of 12.5 to 18.75 metre/second is produced (for 150 km and 100 km, respectively). For a 10 cm radar such as the NEXRAD, the unambiguous velocity range would be doubled.

Some techniques using two alternating pulse repetition frequencies (PRF) allow a greater Doppler range. The velocities noted with the first pulse rate could be equal or different with the second. For instance, if the maximum velocity with a certain rate is 10 metre/second and the one with the other rate is 15 m/s. The data coming from both will be the same up to 10 m/s, and will differ thereafter. It is then possible to find a mathematical relation between the two returns and calculate the real velocity beyond the limitation of the two PRFs.

Doppler Interpretation

In a uniform rainstorm moving eastward, a radar beam pointing west will "see" the raindrops moving toward itself, while a beam pointing east will "see" the drops moving away. When the beam scans to the north or to the south, no relative motion is noted.

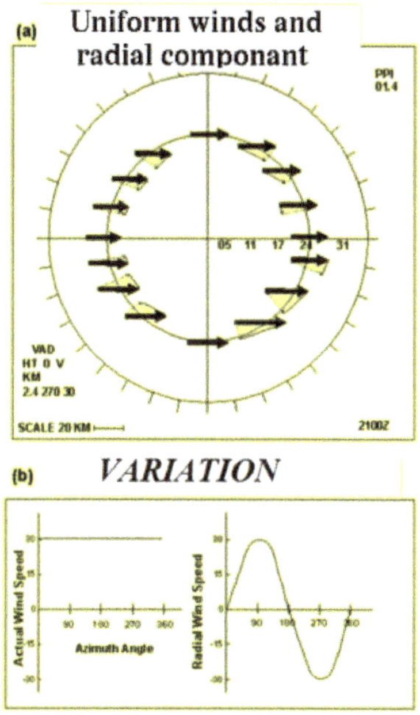

Radial component of real winds when scanning through 360 degrees

Synoptic

In the synoptic scale interpretation, the user can extract the wind at different levels over the radar coverage region. As the beam is scanning 360 degrees around the radar, data will come from all those angles and be the radial projection of the actual wind on the individual angle. The intensity pattern formed by this scan can be represented by a cosine curve (maximum in the precipitation motion and zero in the perpendicular direction). One can then calculate the direction and the strength of the motion of particles as long as there is enough coverage on the radar screen.

However, the rain drops are falling. As the radar only sees the radial component and has a certain elevation from ground, the radial velocities are contaminated by some fraction of the falling speed. This component is negligible in small elevation angles, but must be taken into account for higher scanning angles.

Meso Scale

In the velocity data, there could be smaller zones in the radar coverage where the wind varies from the one mentioned above. For example, a thunderstorm is a mesoscale phenomenon which often includes rotations and turbulence. These may only cover few square kilometers but are visible by variations in the radial speed. Users can recognize velocity patterns in the wind associated with rotations, such as mesocyclone, convergence (outflow boundary) and divergence (downburst).

Polarization

Droplets of falling liquid water tend to have a larger horizontal axis due to the drag coefficient of air while falling (water droplets). This causes the water molecule dipole to be oriented in that direction; so, radar beams are, generally, polarized horizontally in order to receive the maximal signal reflection.

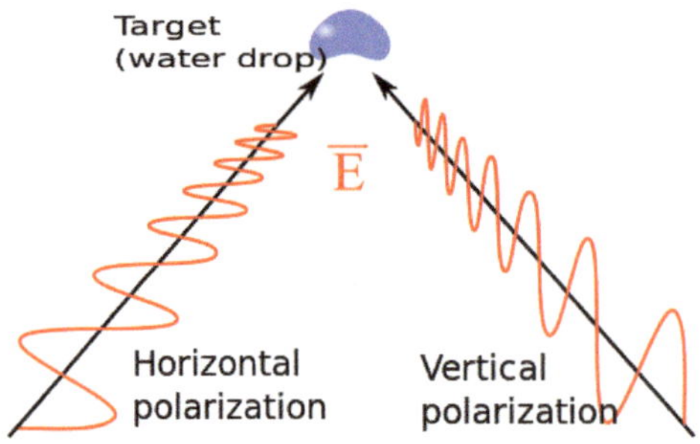

Targeting with dual-polarization will reveal the form of the droplet

If two pulses are sent simultaneously with orthogonal polarization (vertical and horizontal, Z_V and Z_H respectively), two independent sets of data will be received. These signals can be compared in several useful ways:

- Differential Reflectivity (Z_{dr}) – The differential reflectivity is the ratio of the reflected vertical and horizontal power returns as Z_V/Z_H. Among other things, it is a good indicator of drop shape and drop shape is a good estimate of average drop size.
- Correlation Coefficient (ρ_{hv}) – A statistical correlation between the reflected horizontal and vertical power returns. High values, near one, indicate homogeneous precipitation types, while lower values indicate regions of mixed precipitation types, such as rain and snow, or hail, or in extreme cases debris aloft, usually coinciding with a Tornado vortex signature.
- Linear Depolarization Ratio (*LDR*) – This is a ratio of a vertical power return from a horizontal pulse or a horizontal power return from a vertical pulse. It can also indicate regions where there is a mixture of precipitation types.
- Differential Phase (θ_{dp}) – The differential phase is a comparison of the returned phase difference between the horizontal and vertical pulses. This change in phase is caused by the difference in the number of wave cycles (or wavelengths) along the propagation path for horizontal and vertically polarized waves. It should not be confused with the Doppler frequency shift, which is caused by the motion of the cloud and precipitation particles. Unlike the differential reflectivity, correlation coefficient and linear depolarization ratio, which are all dependent on reflected power, the differential phase is a "propagation effect." It is a very good estimator of rain rate and is not affected by attenuation. The range derivative of differential phase (specific differential phase, K_{dp}) can be used to localize areas of strong precipitation/attenuation.

With more information about particle shape, dual-polarization radars can more easily distinguish airborne debris from precipitation, making it easier to locate tornados.

With this new knowledge added to the reflectivity, velocity, and spectrum width produced by Doppler weather radars, researchers have been working on developing algorithms to differentiate precipitation types, non-meteorological targets, and to produce better rainfall accumulation estimates. In the U.S., NCAR and NSSL have been world leaders in this field.

NOAA established a test deployment for dual-polametric radar at NSSL and equipped all its 10 cm NEXRAD radars with dual-polarization, which was completed in April 2013. In 2004, ARMOR

Doppler Weather Radar in Huntsville, Alabama was equipped with a SIGMET Antenna Mounted Receiver, giving Dual-Polarmetric capabilities to the operator. McGill University J. S. Marshall Radar Observatory in Montreal, Canada has converted its instrument (1999) and the data are used operationally by Environment Canada in Montreal. Another Environment Canada radar, in King City (North of Toronto), was dual-polarized in 2005; it uses a 5 cm wavelength, which experiences greater attenuation. Environment Canada is working on converting all of its radars to dual-polarization. Météo-France is planning on incorporating dual-polarizing Doppler radar in its network coverage.

Main Types of Radar Outputs

All data from radar scans are displayed according to the need of the users. Different outputs have been developed through time to reach this. Here is a list of common and specialized outputs available.

Plan Position Indicator

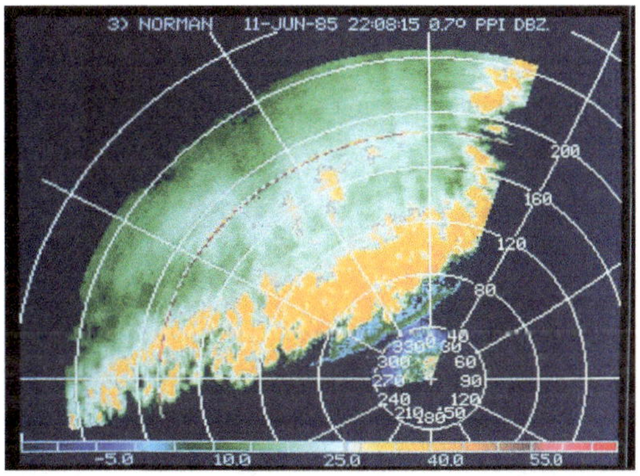

Thunderstorm line viewed in reflectivity (dBZ) on a PPI

Since data are obtained one angle at a time, the first way of displaying them has been the Plan Position Indicator (PPI) which is only the layout of radar return on a two dimensional image. One has to remember that the data coming from different distances to the radar are at different heights above ground.

This is very important as a high rain rate seen near the radar is relatively close to what reaches the ground but what is seen from 160 km away is about 1.5 km above ground and could be far different from the amount reaching the surface. It is thus difficult to compare weather echoes at different distances from the radar.

PPIs are afflicted with ground echoes near the radar as a supplemental problem. These can be misinterpreted as real echoes. So other products and further treatments of data have been developed to supplement such shortcomings.

Usage: Reflectivity, Doppler and polarimetric data can use PPI.

In the case of Doppler data, two points of view are possible: relative to the surface or the storm. When looking at the general motion of the rain to extract wind at different altitudes, it is better to use data relative to the radar. But when looking for rotation or wind shear under a thunderstorm, it is better to use the storm relative images that subtract the general motion of precipitation leaving the user to view the air motion as if he would be sitting on the cloud.

Constant-altitude Plan Position Indicator

To avoid some of the problems on PPIs, the constant-altitude plan position indicator (CAPPI) has been developed by Canadian researchers. It is basically a horizontal cross-section through radar data. This way, one can compare precipitation on an equal footing at difference distance from the radar and avoid ground echoes. Although data are taken at a certain height above ground, a relation can be inferred between ground stations' reports and the radar data.

CAPPIs call for a large number of angles from near the horizontal to near the vertical of the radar to have a cut that is as close as possible at all distance to the height needed. Even then, after a certain distance, there isn't any angle available and the CAPPI becomes the PPI of the lowest angle. The zigzag line on the angles diagram above shows the data used to produce 1.5 km and 4 km height CAPPIs. Notice that the section after 120 km is using the same data.

Usage

Since the CAPPI uses the closest angle to the desired height at each point from the radar, the data can originate from slightly different altitudes, as seen on the image, in different points of the radar coverage. It is therefore crucial to have a large enough number of sounding angles to minimize this height change. Furthermore, the type of data must be changing relatively gradually with height to produce an image that is not noisy.

Reflectivity data being relatively smooth with height, CAPPIs are mostly used for displaying them. Velocity data, on the other hand, can change rapidly in direction with height and CAPPIs of them are not common. It seems that only McGill University is producing regularly Doppler CAPPIs with the 24 angles available on their radar. However, some researchers have published papers using velocity CAPPIs to study tropical cyclones and development of NEXRAD products. Finally, polarimetric data are recent and often noisy. There doesn't seem to have regular use of CAPPI for them although the *SIGMET* company offer a software capable to produce those types of images.

Real-time examples

- McGill University
- Environment Canada

Vertical Composite

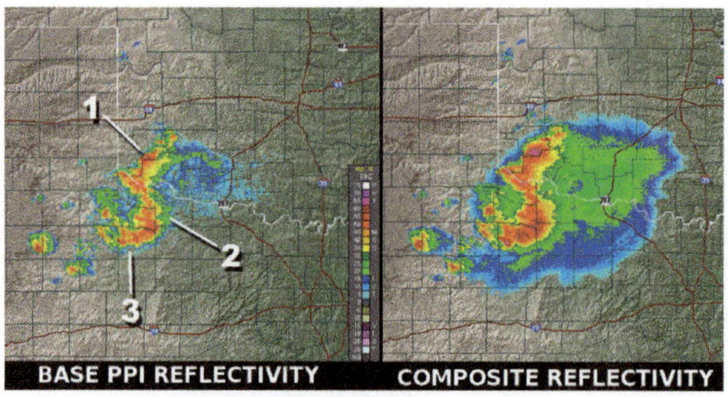

Base PPI versus Composite.

Another solution to the PPI problems is to produce images of the maximum reflectivity in a layer above ground. This solution is usually taken when the number of angles available is small or variable. The American National Weather Service is using such Composite as their scanning scheme can vary from 4 to 14 angles, according to their need, which would make very coarse CAPPIs. The Composite assures that no strong echo is missed in the layer and a treatment using Doppler velocities eliminates the ground echoes. Comparing base and composite products, one can locate virga and updrafts zones.

Real time example: NWS Burlington radar, one can compare the BASE and COMPOSITE products.

Accumulations

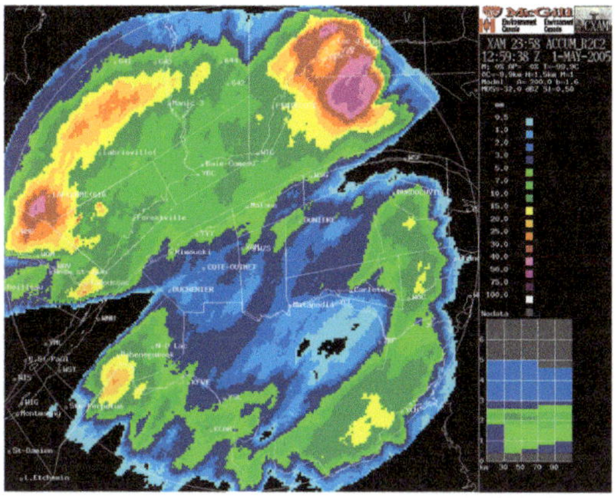

24 hours rain accumulation on the Val d'Irène radar in Eastern Canada. Notice the zones without data in the East and Southwest caused by radar beam blocking from mountains.

Another important use of radar data is the ability to assess the amount of precipitation that has fallen over large basins, to be used in hydrological calculations; such data is useful in flood control, sewer management and dam construction. The computed data from radar weather may be used in conjunction with data from ground stations.

To produce radar accumulations, we have to estimate the rain rate over a point by the average value over that point between one PPI, or CAPPI, and the next; then multiply by the time between those images. If one wants for a longer period of time, one has to add up all the accumulations from images during that time.

Echotops

Aviation is a heavy user of radar data. One map particularly important in this field is the Echotops for flight planning and avoidance of dangerous weather. Most country weather radars are scanning enough angles to have a 3D set of data over the area of coverage. It is relatively easy to estimate the maximum altitude at which precipitation is found within the volume. However, those are not the tops of clouds as they always extend above the precipitation.

Vertical Cross Sections

To know the vertical structure of clouds, in particular thunderstorms or the level of the melting layer, a vertical cross-section product of the radar data is available. This is done by displaying only the data along a line, from coordinates A to B, taken from the different angles scanned.

Range Height Indicator

When a weather radar is scanning in only one direction vertically, it obtains high resolution data along a vertical cut of the atmosphere. The output of this sounding is called a *Range Height Indicator* (RHI) which is excellent for viewing the detailed vertical structure of a storm. This is different from the vertical cross section mentioned above by the fact that the radar is making a vertical cut along specific directions and does not scan over the entire 360 degrees around the site. This kind of sounding and product is only available on research radars.

Radar Networks

Berrimah Radar in Darwin, Northern Territory Australia

Over the past few decades, radar networks have been extended to allow the production of composite views covering large areas. For instance, many countries, including the United States, Canada and much of Europe, produce images that include all of their radars. This is not a trivial task.

In fact, such a network can consist of different types of radar with different characteristics such as beam width, wavelength and calibration. These differences have to be taken into account when matching data across the network, particularly to decide what data to use when two radars cover the same point. If one uses the stronger echo but it comes from the more distant radar, one uses returns that are from higher altitude coming from rain or snow that might evaporate before reaching the ground (virga). If one uses data from the closer radar, it might be attenuated passing through a thunderstorm. Composite images of precipitations using a network of radars are made with all those limitations in mind.

Automatic Algorithms

To help meteorologists spot dangerous weather, mathematical algorithms have been introduced in the weather radar treatment programs. These are particularly important in analyzing the Doppler velocity data as they are more complex. The polarization data will even need more algorithms.

Main algorithms for reflectivity:

- Vertically Integrated Liquid (VIL) is an estimate of the total mass of precipitation in the clouds.
- *VIL Density* is VIL divided by the height of the cloud top. It is a clue to the possibility of large hail in thunderstorms.
- *Potential wind gust*, which can estimate the winds under a cloud (a downdraft) using the VIL and the height of the echotops (radar estimated top of the cloud) for a given storm cell.
- Hail algorithms that estimate the presence of hail and its probable size.

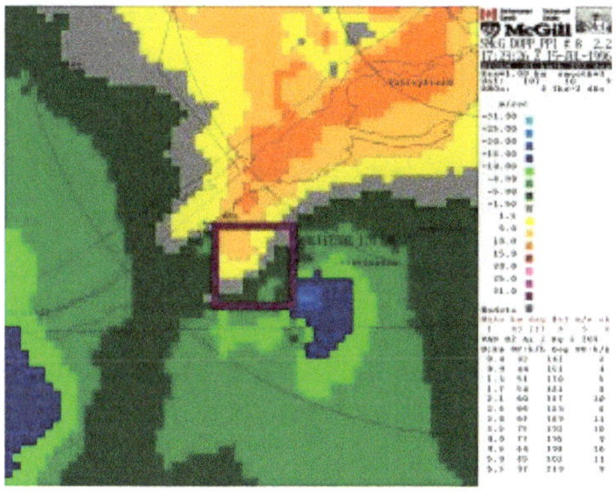

The square in this Doppler image has been automatically placed by the radar program to spot the position of a mesocyclone. Notice the inbound/outbound doublet (blue/yellow) with the zero velocity line (gray) parallel to the radial to the radar (up right). It is noteworthy to mention that the change in wind direction here occurs over less than 10 km.

Main algorithms for Doppler velocities:

- Mesocyclone detection: it is triggered by a velocity change over a small circular area. The algorithm is searching for a "*doublet*" of inbound/outbound velocities with the zero line of velocities, between the two, along a radial line from the radar. Usually the mesocyclone detection must be found on two or more stacked progressive tilts of the beam to be significative of rotation into a thunderstorm cloud.
- TVS or Tornado Vortex Signature algorithm is essentially a mesocyclone with a large velocity threshold found through many scanning angles. This algorithm is used in NEXRAD to indicate the possibility of a tornado formation.
- Wind shear in low levels. This algorithm detects variation of wind velocities from point to point in the data and looking for a *doublet* of inbound/outbound velocities with the zero line perpendicular to the radar beam. The wind shear is associated with downdraft, (downburst and microburst), gust fronts and turbulence under thunderstorms.
- VAD Wind Profile (VWP) is a display that estimates the direction and speed of the horizontal wind at various upper levels of the atmosphere, using the technique explained in the Doppler section.

Radar Integrated Display with Geospatial Elements

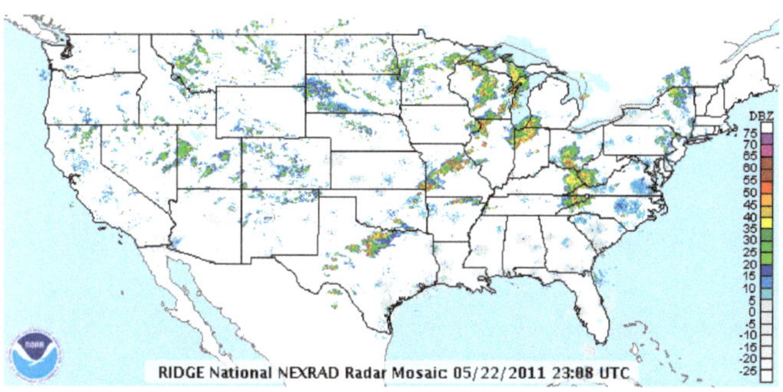

Map of the RIDGE presentation of 2011 Joplin tornado.

A new popular presentation of weather radar data in United States is via *Radar Integrated Dis-*

play with Geospatial Elements (RIDGE) in which the radar data is projected on a map with geospatial elements such as topography maps, highways, state/county boundaries and weather warnings. The projection often is flexible giving the user a choice of various geographic elements. It is frequently used in conjunction with animations of radar data over a time period.

Limitations and Artifacts

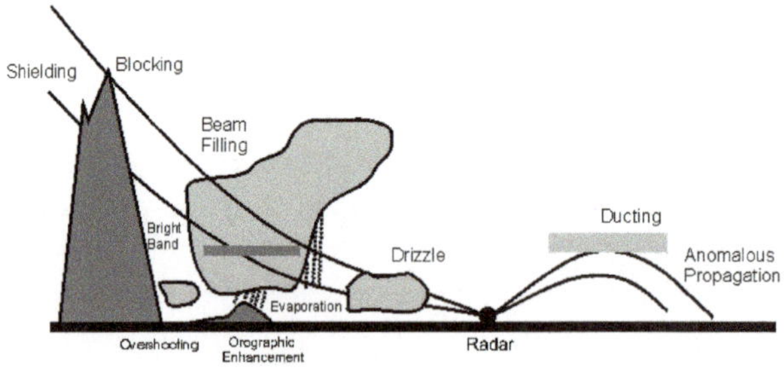

Radar data interpretation depends on many hypotheses about the atmosphere and the weather targets, including:

- International Standard Atmosphere.
- Targets small enough to obey the Rayleigh scattering, resulting in the return being proportional to the precipitation rate.
- The volume scanned by the beam is full of *meteorological* targets (rain, snow, etc.), all of the same variety and in a uniform concentration.
- No attenuation
- No amplification
- Return from side lobes of the beam are negligible.
- The beam is close to a Gaussian function curve with power decreasing to half at half the width.
- The outgoing and returning waves are similarly polarized.
- There is no return from multiple reflections.

These assumptions are not always met; one must be able to differentiate between reliable and dubious echoes.

Anomalous Propagation (Non-standard Atmosphere)

The first assumption is that the radar beam is moving through air that cools down at a certain rate with height. The position of the echoes depend heavily on this hypothesis. However, the real atmosphere can vary greatly from the norm.

Super Refraction

Temperature inversions often form near the ground, for instance by air cooling at night while remaining warm aloft. As the index of refraction of air decreases faster than normal the radar beam bends toward the ground instead of continuing upward. Eventually, it will hit the ground and be reflected back toward the radar. The processing program will then wrongly place the return echoes at the height and distance it would have been in normal conditions.

This type of false return is relatively easy to spot on a time loop if it is due to night cooling or marine inversion as one sees very strong echoes developing over an area, spreading in size laterally but not moving and varying greatly in intensity. However, inversion of temperature exists ahead of warm fronts and the abnormal propagation echoes are then mixed with real rain.

The extreme of this problem is when the inversion is very strong and shallow, the radar beam reflects many times toward the ground as it has to follow a waveguide path. This will create multiple bands of strong echoes on the radar images.

This situation can be found with inversions of temperature aloft or rapid decrease of moisture with height. In the former case, it could be difficult to notice.

Under Refraction

On the other hand, if the air is unstable and cools faster than the standard atmosphere with height, the beam ends up higher than expected. This indicates that precipitation is occurring higher than the actual height. Such an error is difficult to detect without additional temperature lapse-rate data for the area.

Non-Rayleigh Targets

If we want to reliably estimate the precipitation rate, the targets have to be 10 times smaller than the radar wave according to Rayleigh scattering. This is because the water molecule has to be excited by the radar wave to give a return. This is relatively true for rain or snow as 5 or 10 cm wavelength radars are usually employed.

However, for very large hydrometeors, since the wavelength is on the order of stone, the return levels off according to Mie theory. A return of more than 55 dBZ is likely to come from hail but won't vary proportionally to the size. On the other hand, very small targets such as cloud droplets are too small to be excited and do not give a recordable return on common weather radars.

Resolution and Partially Filled Scanned Volume

As demonstrated at the start of the article, radar beams have a physical dimension and data are sampled at discrete angles, not continuously, along each angle of elevation. This results in an averaging of the values of the returns for reflectivity, velocities and polarization data on the resolution volume scanned.

In the figure to the left, at the top is a view of a thunderstorm taken by a wind profiler as it was passing overhead. This is like a vertical cross section through the cloud with 150-metre vertical and 30-metre horizontal resolution. The reflectivity has large variations in a short distance. Compare this with a simulated view of what a regular weather radar would see at 60 km, in the bottom of the figure. Everything has been smoothed out. Not only the coarser resolution of the radar blur the image but the sounding incorporates area that are echo free, thus extending the thunderstorm beyond its real boundaries.

PARTIAL BEAMFILLING ON REFLECTIVITY GRADIENT

Thunderstorm cross-section from a wind profiler

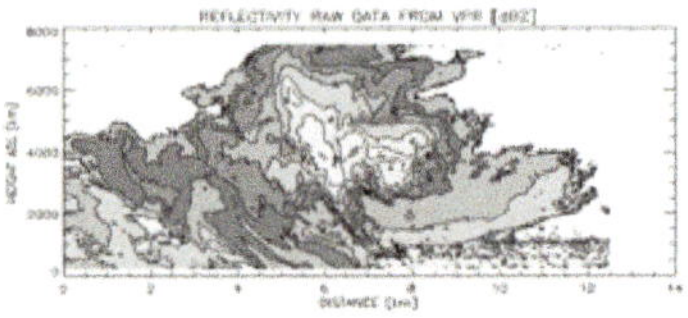

Simulation of the same data but with a scanning radar at 60 km distance

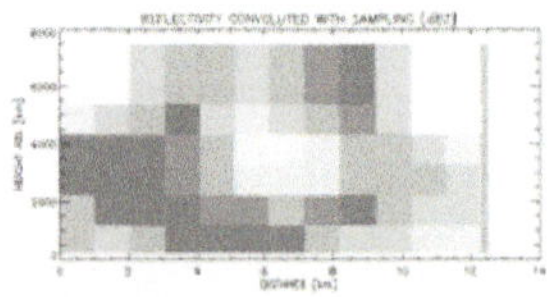

Profiler high resolution view of a thunderstorm (top) and by a weather radar (bottom).

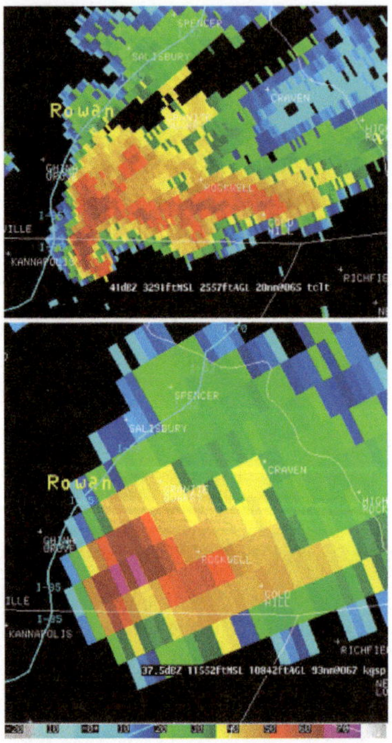

A supercell thunderstorm seen from two radars almost colocated.
The top image is from a TDWR and the bottom one from a NEXRAD.

This shows how the output of weather radar is only an approximation of reality. The image to the right compares real data from two radars almost colocated. The TDWR has about half the beam-width of the other and one can see twice more details than with the NEXRAD.

Resolution can be improved by newer equipment but some things cannot. As mentioned previously, the volume scanned increases with distance so the possibility that the beam is only partially filled also increases. This leads to underestimation of the precipitation rate at larger distances and fools the user into thinking that rain is lighter as it moves away.

Beam Geometry

The radar beam has a distribution of energy similar to the diffraction pattern of a light passing through a slit. This is because the wave is transmitted to the parabolic antenna through a slit in the wave-guide at the focal point. Most of the energy is at the center of the beam and decreases along a curve close to a Gaussian function on each side. However, there are secondary peaks of emission that will sample the targets at off-angles from the center. Designers attempt to minimize the power transmitted by such lobes, but they cannot be completely eliminated.

When a secondary lobe hits a reflective target such as a mountain or a strong thunderstorm, some of the energy is reflected to the radar. This energy is relatively weak but arrives at the same time that the central peak is illuminating a different azimuth. The echo is thus misplaced by the processing program. This has the effect of actually broadening the real weather echo making a smearing of weaker values on each side of it. This causes the user to overestimate the extent of the real echoes.

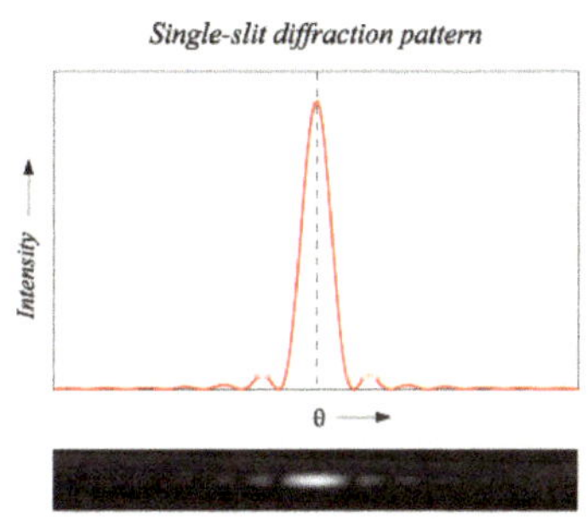

Idealized energy distribution of a radar beam (Central lobe at 0 and secondary lobes on each side)

Diffraction by a circular slit simulating the energy viewed by weather targets

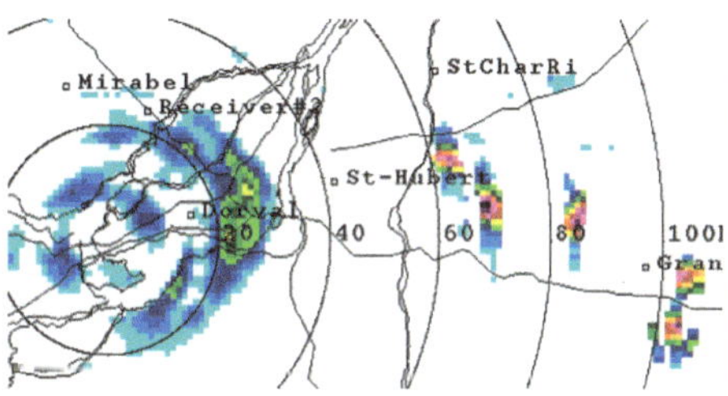

The strong echoes are returns of the central peak of the radar from a series of small hills (yellow and reds pixels). The weaker echoes on each sides of them are from secondary lobes (blue and green)

Non-weather Targets

There is more than rain and snow in the sky. Other objects can be misinterpreted as rain or snow by weather radars. Insects and arthropods are swept along by the prevailing winds, while birds follow their own course. As such, fine line patterns within weather radar imagery, associated with converging winds, are dominated by insect returns. Bird migration, which tends to occur overnight within the lowest 2000 metres of the Earth's atmosphere, contaminates wind profiles gathered by weather radar, particularly the WSR-88D, by increasing the environmental wind returns by 30–60 km/hr. Other objects within radar imagery include:

- Thin metal strips (chaff) dropped by military aircraft to fool enemies.
- Solid obstacles such as mountains, buildings, and aircraft.
- Ground and sea clutter.
- Reflections from nearby buildings ("urban spikes").

Such extraneous objects have characteristics that allow a trained eye to distinguish them. It is also possible to eliminate some of them with post-treatment of data using reflectivity, Doppler, and polarization data.

Wind Farms

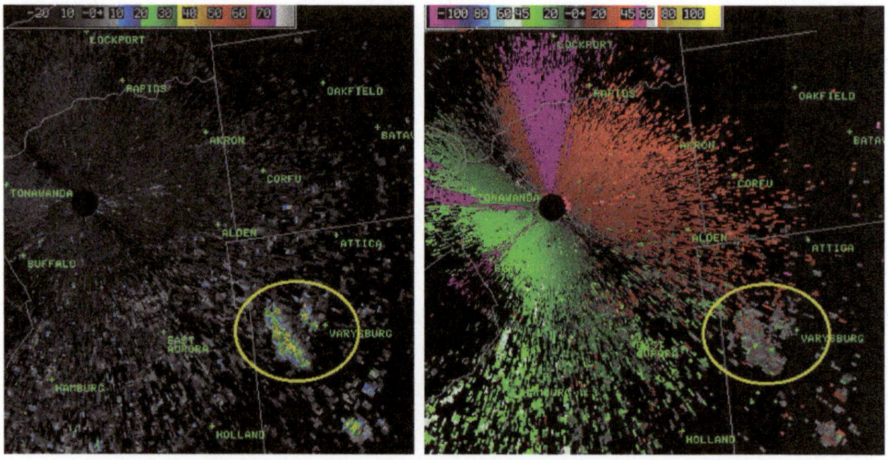

Reflectivity (left) and radial velocities (right) southeast of a NEXRAD weather radar. Echoes in circles are from a wind farm.

The rotating blades of windmills on modern wind farms can return the radar beam to the radar if they are in its path. Since the blades are moving, the echoes will have a velocity and can be mistaken for real precipitation. The closer the wind farm, the stronger the return, and the combined signal from many towers is stronger. In some conditions, the radar can even see toward and away velocities that generate false positives for the tornado vortex signature algorithm on weather radar; such an event occurred in 2009 in Dodge City, Kansas.

As with other structures that stand in the beam, attenuation of radar returns from beyond windmills may also lead to underestimation.

Attenuation

Microwaves used in weather radars can be absorbed by rain, depending on the wavelength used. For 10 cm radars, this attenuation is negligible. That is the reason why countries with high water content storms are using 10 cm wavelength, for example the US NEXRAD. The cost of a larger antenna, klystron and other related equipment is offset by this benefit.

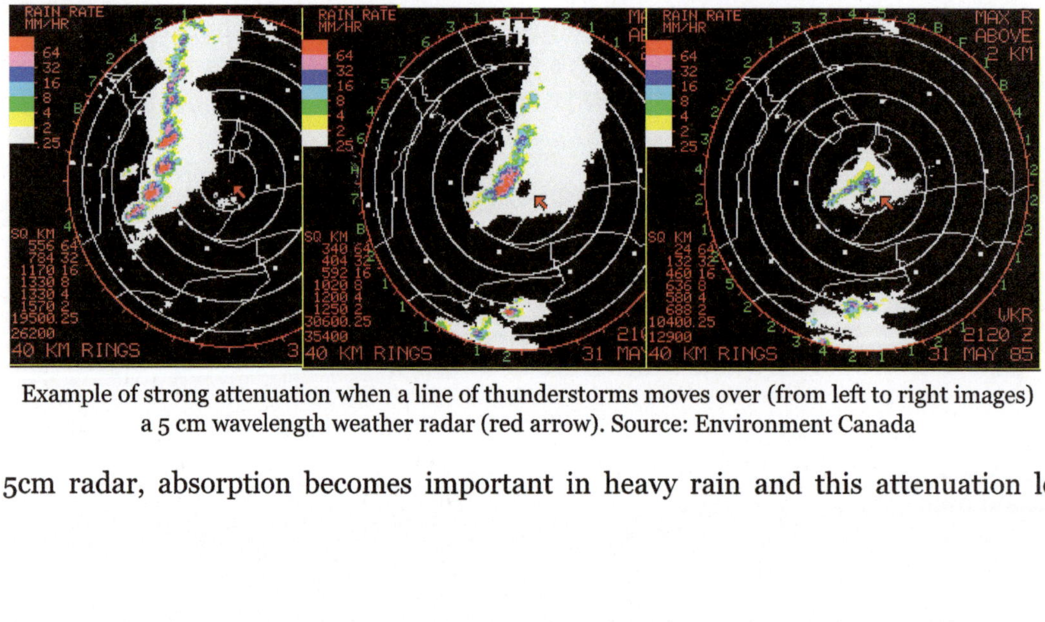

Example of strong attenuation when a line of thunderstorms moves over (from left to right images) a 5 cm wavelength weather radar (red arrow). Source: Environment Canada

For a 5cm radar, absorption becomes important in heavy rain and this attenuation leads to underestimation of echoes in and beyond a strong thunderstorm. Canada and other northern countries use this less costly kind of radar as the precipitation in such areas is usually less intense. However, users must consider this characteristic when interpreting data. The images above show how a strong line of echoes seems to vanish as it moves over the radar. To compensate for this behaviour, radar sites are often chosen to somewhat overlap in coverage to give different points of view of the same storms.

Shorter wavelengths are even more attenuated and are only useful on short range radar. Many television stations in the United States have 3 cm radars to cover their audience area. Knowing their limitations and using them with the local NEXRAD can supplement the data available to a meteorologist.

Due to the spread of dual-polarization radar systems, robust and efficient approaches for the compesantion of rain attenuation are currently implemented by operational weather services.

Bright Band

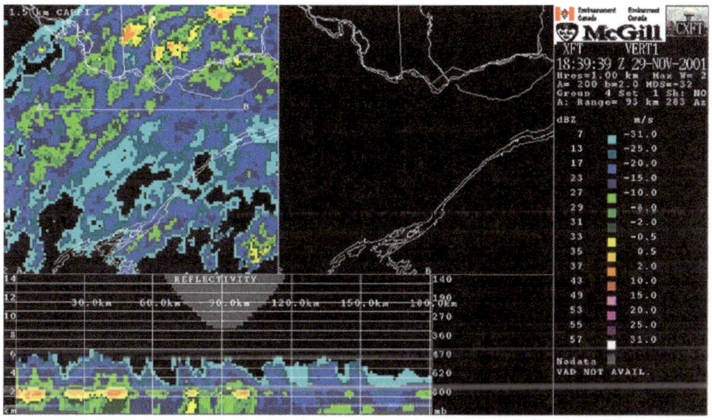

1.5 km altitude CAPPI at the top with strong contamination from the brightband (yellows). The vertical cut at the bottom shows that this strong return is only above ground.

A radar beam's reflectivity depends on the diameter of the target and its capacity to reflect. Snowflakes are large but weakly reflective while rain drops are small but highly reflective.

When snow falls through a layer above freezing temperature, it melts into rain. Using the reflectivity equation, one can demonstrate that the returns from the snow before melting and the rain after, are not too different as the change in dielectric constant compensates for the change in size. However, during the melting process, the radar wave "sees" something akin to very large droplets as snow flakes become coated with water.

This gives enhanced returns that can be mistaken for stronger precipitations. On a PPI, this will show up as an intense ring of precipitation at the altitude where the beam crosses the melting level while on a series of CAPPIs, only the ones near that level will have stronger echoes. A good way to confirm a bright band is to make a vertical cross section through the data, as illustrated in the picture above.

An opposite problem is that drizzle (precipitation with small water droplet diameter) tends not to show up on radar because radar returns are proportional to the sixth power of droplet diameter.

Multiple Reflections

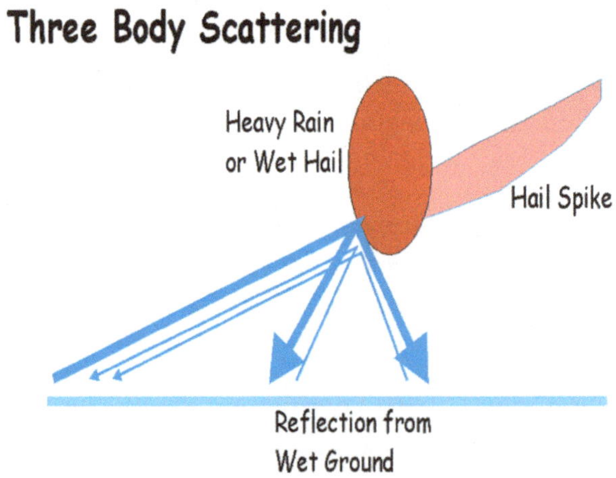

It is assumed that the beam hits the weather targets and returns directly to the radar. In fact, there is energy reflected in all directions. Most of it is weak, and multiple reflections diminish it even further so what can eventually return to the radar from such an event is negligible. However, some situations allow a multiple-reflected radar beam to be received by the radar antenna. For instance, when the beam hits hail, the energy spread toward the wet ground will be reflected back to the hail and then to the radar. The resulting echo is weak but noticeable. Due to the extra path length it has to go through, it arrives later at the antenna and is placed further than its source. This gives a kind of triangle of false weaker reflections placed radially behind the hail.

Solutions for Now and the Future

Filtering

These two images show what can be presently achieved to clean up radar data. The output on the left is made with the raw returns and it is difficult to spot the real weather. Since rain and snow clouds are usually moving, one can use the Doppler velocities to eliminate a good part of the clutter (ground echoes, reflections from buildings seen as urban spikes, anomalous propagation). The image on the right has been filtered using this property.

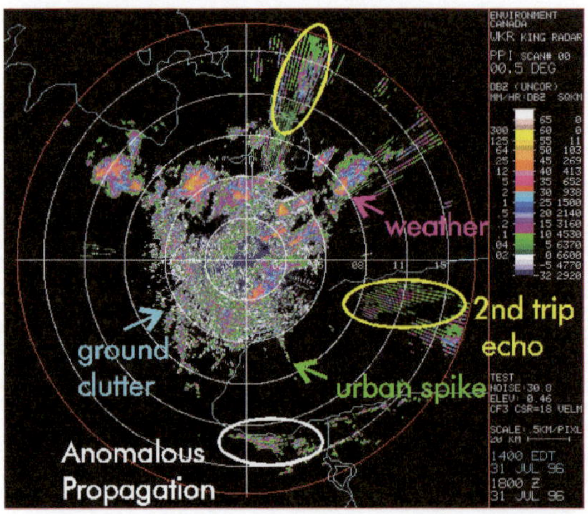

Radar image of reflectivity with many non-weather echoes.

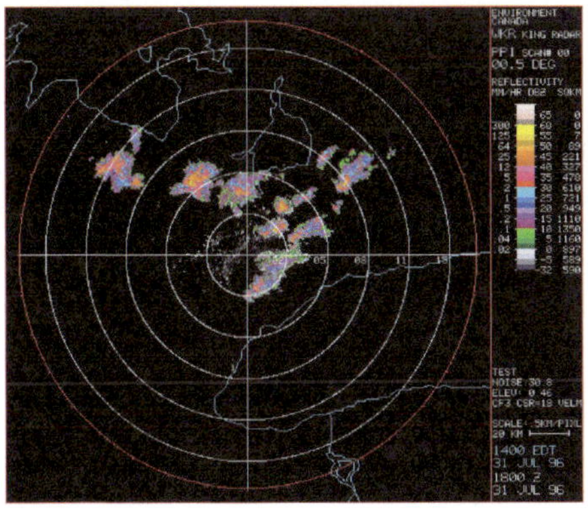

The same image but cleaned using the Doppler velocities.

However, not all non-meteorological targets remain still (birds, insects, dust). Others, like the bright band, depend on the structure of the precipitation. Polarization offers a direct typing of the echoes which could be used to filter more false data or produce separate images for specialized purposes. This recent development is expected to improve the quality of radar products.

Mesonet

Phased Array Weather Radar in Norman, Oklahoma

Another question is the resolution. As mentioned previously, radar data are an average of the scanned volume by the beam. Resolution can be improved by larger antenna or denser networks. A program by the Center for Collaborative Adaptive Sensing of the Atmosphere (CASA) aims to supplement the regular NEXRAD (a network in the United States) using many low cost X-band (3 cm) weather radar mounted on cellular telephone towers. These radars will subdivide the large area of the NEXRAD into smaller domains to look at altitudes below its lowest angle. These will give details not currently available.

Using 3 cm radars, the antenna of each radar is small (about 1 meter diameter) but the resolution is similar at short distance to that of NEXRAD. The attenuation is significant due to the wavelength used but each point in the coverage area is seen by many radars, each viewing from a different direction and compensating for data lost from others.

Scanning Strategies

The number of elevation scanned and the time taken for a complete cycle depend on the weather situation. For instance, with little or no precipitation, the scheme may be limited the lowest angles and using longer impulses in order to detect wind shift near the surface. On the other hand, in violent thundertorm situations, it is better to scan on a large number of angles in order to have a 3 dimensions view of the precipitations as often as possible. To mitigate those different demands, scanning strategies have been developed according to the type of radar, the wavelength used and the most commons weather situations in the area considered.

One example of scanning strategies is given by the US NEXRAD radar network which has evolved with time. For instance, in 2008, it added extra resolution of data, and in 2014, additional intra-cycle scanning of lowest level elevation (MESO-SAILS).

Electronic Sounding

Timeliness is also a point needing improvement. With 5 to 10 minutes time between complete scans of weather radar, much data is lost as a thunderstorm develops. A Phased-array radar is being tested at the National Severe Storms Lab in Norman, Oklahoma, to speed the data gathering. A team in Japan has deployed a phased-array radar for 3D NowCasting at the RIKEN Advanced Institute for Computational Science (AICS).

Specialized Applications

Weather radar on the wing of Pilatus PC-12

Avionics Weather Radar

Aircraft application of radar systems include weather radar, collision avoidance, target tracking, ground proximity, and other systems. For commercial weather radar, ARINC 708 is the primary specification for weather radar systems using an airborne pulse-Doppler radar.

Antennas

Unlike ground weather radar, which is set at a fixed angle, airborne weather radar is being utilized from the nose or wing of an aircraft. Not only will the aircraft be moving up, down, left, and right, but it will be rolling as well. To compensate for this, the antenna is linked and calibrated to the vertical gyro located on the aircraft. By doing this, the pilot is able to set a pitch or angle to the antenna that will enable the stabilizer to keep the antenna pointed in the right direction under moderate maneuvers. The small servo motors will not be able to keep up with abrupt maneuvers, but it will try. In doing this the pilot is able to adjust the radar so that it will point towards the weather system of interest. If the airplane is at a low altitude, the pilot would want to set the radar above the horizon line so that ground clutter is minimized on the display. If the airplane is at a very high altitude, the pilot will set the radar at a low or negative angle, to point the radar towards the clouds wherever they may be relative to the aircraft. If the airplane changes attitude, the stabilizer will adjust itself accordingly so that the pilot doesn't have to fly with one hand and adjust the radar with the other.

Receivers/Transmitters

There are two major systems when talking about the receiver/transmitter: the first is high-powered systems, and the second is low-powered systems; both of which operate in the X-band frequency range (8,000 – 12,500 MHz). High-powered systems operate at 10,000 – 60,000 watts. These systems consist of magnetrons that are fairly expensive (approximately $1,700) and allow for considerable noise due to irregularities with the system. Thus, these systems are highly dangerous for arcing and are not safe to be used around ground personnel. However, the alternative would be the low-powered systems. These systems operate 100 – 200 watts, and require a combination of high gain receivers, signal microprocessors, and transistors to operate as effectively as the high-powered systems. The complex microprocessors help to eliminate noise, providing a more accurate and detailed depiction of the sky. Also, since there are fewer irregularities throughout the system, the low-powered radars can be used to detect turbulence via the Doppler Effect. Since low-powered systems operate at considerable less wattage, they are safe from arcing and can be used at virtually all times.

Thunderstorm Tracking

Digital radar systems now have capabilities far beyond that of their predecessors. Digital systems now offer thunderstorm tracking surveillance. This provides users with the ability to acquire detailed information of each storm cloud being tracked. Thunderstorms are first identified by matching precipitation raw data received from the radar pulse to some sort of template preprogrammed into the system. In order for a thunderstorm to be identified, it has to meet strict definitions of intensity and shape that set it apart from any non-convective cloud. Usually, it must show signs of organization in the horizontal and continuity in the vertical: a core or a more intense center to be identified and tracked by digital radar tracking systems. Once the thunderstorm cell is identified, speed, distance covered, direction, and Estimated Time of Arrival (ETA) are all tracked and recorded to be utilized later.

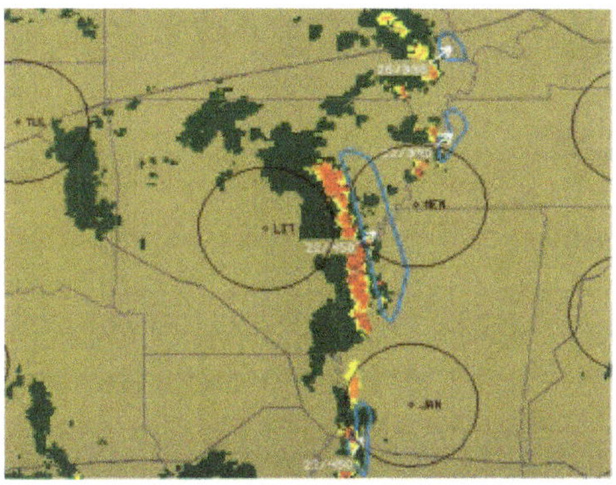

Nowcasting a line of thunderstorms from *AutoNowcaster* system

Doppler Radar and Bird Migration

Using the Doppler weather radar is not limited to determine the location and velocity of precipitation, but it can track bird migrations as well as seen in the non-weather targets section. The radio waves sent out by the radars bounce off rain and birds alike (or even insects like butterflies). The US *National Weather Service*, for instance, have reported having the flights of birds appear on their radars as clouds and then fade away when the birds land. The U.S. National Weather Service St. Louis has even reported monarch butterflies appearing on their radars.

Different programs in North America use regular weather radars and specialized radar data to determine the paths, height of flight, and timing of migrations. This is useful information in planning for windmill farms placement and operation, to reduce bird fatalities, aviation safety and other wildlife management. In Europe, there has been similar developments and even a comprehensive forecast program for aviation safety, based on radar detection.

Meteorite Fall Detection

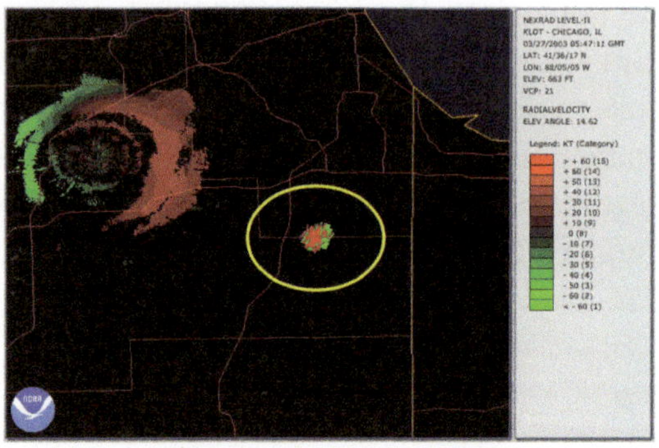

NOAA NEXRAD radar image of the Park Forest, IL, meteorite fall of 26 March 2003.

According to the American Meteor Society, meteorite falls occur on a daily basis somewhere on Earth. However, the database of worldwide meteorite falls maintained by the Meteoritical Society typically records only about 10-15 new meteorite falls annually.

Meteorite occur when a meteoroid falls into the Earth's atmosphere, generating an optically bright meteor by ionization and frictional heating. If the meteoroid is large enough and infall velocity is low enough, surviving meteorites will reach the ground. When the falling meteorites decelerate below about 2–4 km/s, usually at an altitude between 15 and 25 km, they no longer generate an optically bright meteor and enter "dark flight". Because of this, most meteorite falls occurring into the oceans, during the day, or otherwise go unnoticed.

It is in dark flight that falling meteorites typically fall through the interaction volumes of most types of radars. It as been demonstrated that it is possible to identify falling meteorites in weather radar imagery by different studies. This is especially useful for meteorite recovery, as weather radar are part of widespread networks and scan the atmosphere continuously. Furthermore, the meteorites are cause a perturbance of local winds by turbulence, which is noticeable on Doppler outputs, and are falling nearly vertically so their resting place on the ground is close to their radar signature.

An example is shown in the image on the right, showing the Park Forest, IL meteorite fall which occurred on 26 March 2003. The red/green feature at upper left is the motion of clouds near the radar itself, and a signature of falling meteorites is seen inside the yellow ellipse at image center. The intermixed red and green pixels indicate turbulence, in this case arising from the wakes of falling, high-velocity meteorites.

Wind Profiler

WindCollector2 with a Scintec SODAR SFAS

A wind profiler is a type of weather observing equipment that uses radar or sound waves (SODAR) to detect the wind speed and direction at various elevations above the ground. Readings are made at each kilometer above sea level, up to the extent of the troposphere (i.e., between 8 and 17 km above mean sea level). Above this level there is inadequate water vapor present to produce a radar "bounce." The data synthesized from wind direction and speed is very useful to meteorological forecasting and timely reporting for flight planning. A twelve-hour history of data is available through NOAA websites.

Principle

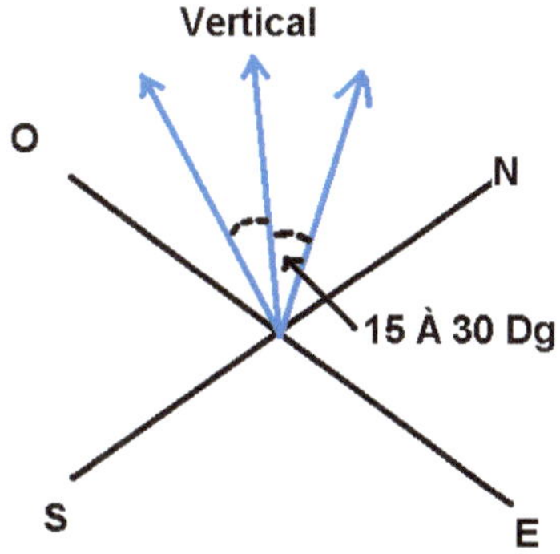

Orientation of the beams in the case of a three tilted wind profiler

In a typical implementation, the radar or sodar can sample along each of five beams: one is aimed vertically to measure vertical velocity, and four are tilted off vertical and oriented orthogonal to one another to measure the horizontal components of the air's motion. A profiler's ability to measure winds is based on the assumption that the turbulent eddies that induce scattering are carried along by the mean wind. The energy scattered by these eddies and received by the profiler is orders of magnitude smaller than the energy transmitted. However, if sufficient samples can be obtained, then the amplitude of the energy scattered by these eddies can be clearly identified above the background noise level, then the mean wind speed and direction within the volume being sampled can be determined. The radial components measured by the tilted beams are the vector sum of the horizontal motion of the air toward or away from the radar and any vertical motion present in the beam. Using appropriate trigonometry, the three-dimensional meteorological velocity components (u,v,w) and wind speed and wind direction are calculated from the radial velocities with corrections for vertical motions.

Radar Wind Profiler

A radar wind profiler.

Pulse-Doppler radar wind profilers operate using electromagnetic (EM) signals to remotely sense winds aloft. The radar transmits an electromagnetic pulse along each of the antenna's pointing directions. A UHF profiler includes subsystems to control the radar's transmitter, receiver, signal processing, and Radio Acoustic Sounding System (RASS), if provided, as well as data telemetry and remote control.

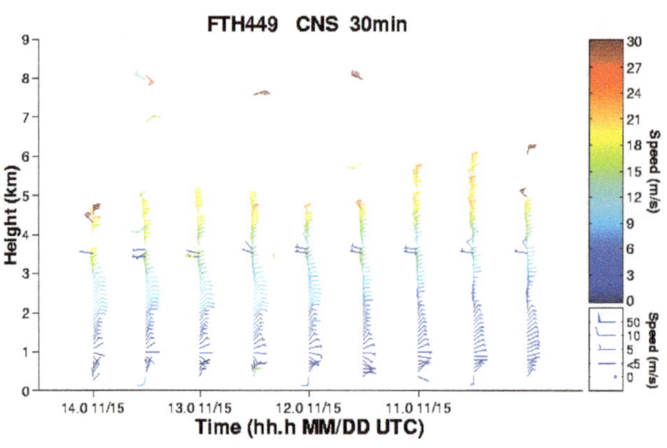

Horizontal plotted wind from a profiler.

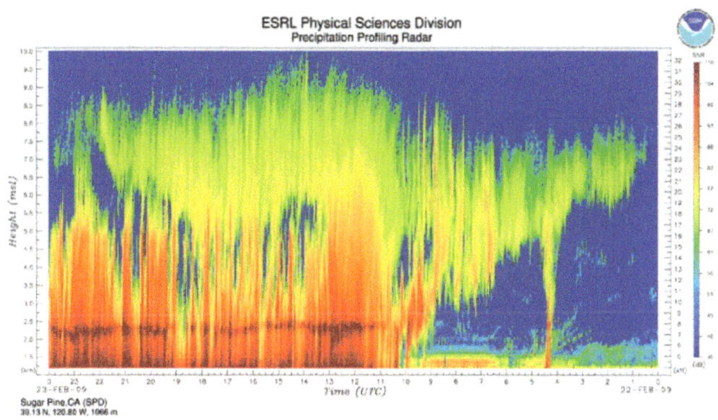

Reflectivity data obtained as a byproduct on a typical radar wind profiler.

The duration of the transmission determines the length of the pulse emitted by the antenna, which in turn corresponds to the volume of air illuminated (in electrical terms) by the radar beam. Small amounts of the transmitted energy are scattered back (referred to as backscattering) toward and received by the radar. Delays of fixed intervals are built into the data processing system so that the radar receives scattered energy from discrete altitudes, referred to as range gates. The Doppler frequency shift of the backscattered energy is determined, and then used to calculate the velocity of the air toward or away from the radar along each beam as a function of altitude. The source of the backscattered energy (radar "targets") is small-scale turbulent fluctuations that induce irregularities in the radio refractive index of the atmosphere. The radar is most sensitive to scattering by turbulent eddies whose spatial scale is ½ the wavelength of the radar, or approximately 16 centimeters (cm) for a UHF profiler.

A boundary-layer radar wind profiler can be configured to compute averaged wind profiles for periods ranging from a few minutes to an hour. Boundary-layer radar wind profilers are often configured to sample in more than one mode. For example, in a "low mode," the pulse of energy transmitted by the profiler may be 60 m in length. The pulse length determines the depth of the column of air being sampled and thus the vertical resolution of the data. In a "high mode," the pulse length is increased, usually to 100 m or greater. The longer pulse length means that more energy is being transmitted for each sample, which improves the signal-to-noise ratio (SNR) of the data. Using a longer pulse length increases the depth of the sample volume and thus decreases the vertical resolution in the data. The greater energy output of the high mode increases the maximum altitude to which the radar wind profiler can sample, but at the expense of coarser vertical resolution and an increase in the altitude at which the first winds are measured. When radar wind profilers are operated in multiple modes, the data are often combined into a single overlapping data set to simplify postprocessing and data validation procedures.

Sodar Wind Profiler

Alternatively, a wind profiler may use sound waves to measure wind speed at various heights above the ground, and the thermodynamic structure of the lower layer of the atmosphere. These sodars can be divided in mono-static system using the same antenna for transmitting and receiving, and bi-static system using separate antennas. The difference between the two antenna systems determines whether atmospheric scattering is by temperature fluctuations (in mono-static systems), or by both temperature and wind velocity fluctuations (in bi-static systems).

The TRITON transportable SODAR system used to measure wind profiles from Second Wind.

Mono-static antenna systems can be divided further into two categories: those using multiple axis, individual antennas and those using a single phased array antenna. The multiple-axis systems generally use three individual antennas aimed in specific directions to steer the acoustic beam. One antenna is generally aimed vertically, and the other two are tilted slightly from the vertical at an orthogonal angle. Each of the individual antennas may use a single transducer focused into a parabolic reflector to form a parabolic loudspeaker, or an array of speaker drivers and horns (transducers) all transmitting in-phase to form a single beam. Both the tilt angle from the vertical and the azimuth angle of each antenna are fixed when the system is set up.

The vertical range of sodars is approximately 0.2 to 2 kilometers (km) and is a function of frequency, power output, atmospheric stability, turbulence, and, most importantly, the noise environment in which a sodar is operated. Operating frequencies range from less than 1000 Hz to over 4000 Hz, with power levels up to several hundred watts. Due to the attenuation characteristics of the atmosphere, high power, lower frequency sodars will generally produce greater height coverage. Some sodars can be operated in different modes to better match vertical resolution and range to the application. This is accomplished through a relaxation between pulse length and maximum altitude.

Hail Spike

A hail spike or three body scatter spike (TBSS) is an artifact on a weather radar display indicative of large hail. They are identified by a spike of weak reflectivity echoes that extend out from a thunderstorm, and away from the radar site.

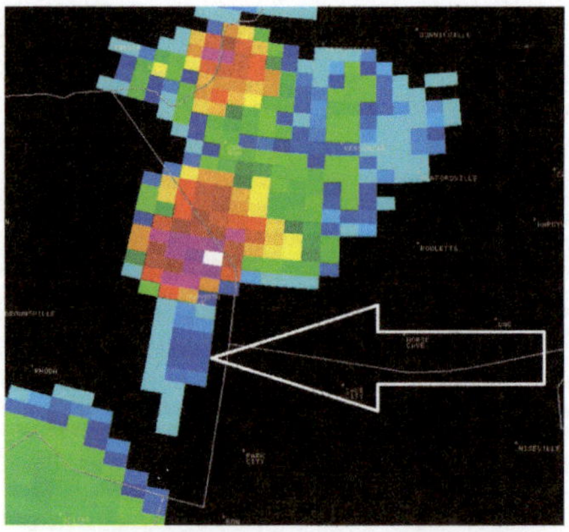

Example of a three body spike : the weak triangular echoes behind the red and white thunderstorm core (pointed by an arrow).

Cause

Generally known as hail spikes, these are the result of energy from the radar hitting hail, or very heavy rain, and being reflected to the ground, where they reflect back to the hail and then to the radar as in the image on the left. This results in the radar picking up the energy from the multiple path at a later time than the energy that came back directly from the hail to the radar. Both are however on the same radial angle from the radar as the antenna did not have the time to turn significantly.

The multipath echoes are then analyzed on the radar display as echoes extending in a radial direction behind the actual location of the hail/heavy rain core. The loss of energy by due to multiple reflections means weaker return echoes. The hail spike region has thus comparatively quite weaker echoes than the echoes directly from the hail or heavy rain core.

Since hail cores are most intense at higher elevations, hail spikes usually appear at the levels aloft that accompany the most intense hail. Because of this, hail spikes are usually not seen at lower elevations. Another restriction to detection is that the signal of the radar beam has to do multiple reflections, each time weakening it. So hail spikes are usually noticeable only in extremely large hailstone cases.

Multiple TBSS Signatures

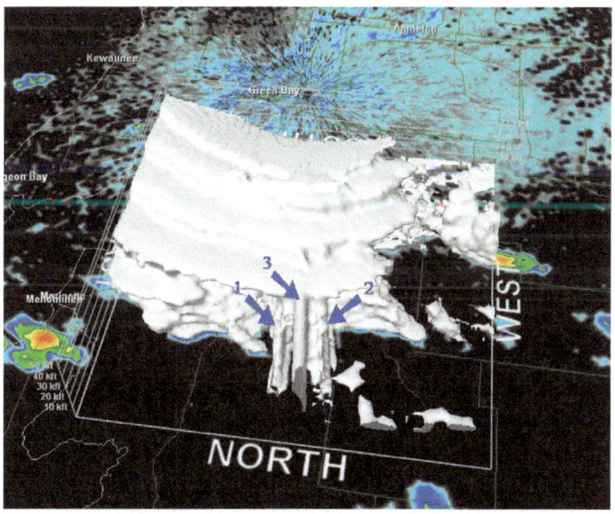

3-D image showing three separate TBSS signatures in one volume scan

In rare cases more than one hail spike has been documented with a single storm in one volume scan. This indicates that more than one hail core exists in the same storm and far enough apart that each core can be sampled by the radar individually.

Use in Forecasting

Because of their observed accuracy in indicating large hail aloft, TBSS's are used operationally by the National Weather Service to identify thunderstorms that could likely produce large, severe hail. This would warrant the issuance of a severe thunderstorm warning or mention of large hail in a tornado warning.

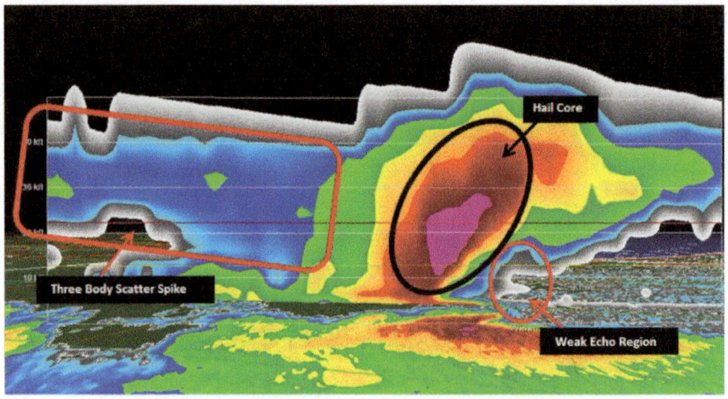

A cross-Section of a Three body spike from an Anti-Cyclonic Thunderstorm.
The spike is the boxed areas to the left.

Terminal Doppler Weather Radar

Airports with a TDWR in the US. Another in San Juan, Puerto Rico, is not on this map.

Terminal Doppler Weather Radar (TDWR) is a Doppler weather radar system with a three-dimensional "pencil beam" used primarily for the detection of hazardous wind shear conditions, precipitation, and winds aloft on and near major airports situated in climates with great exposure to thunderstorms in the United States. As of 2011, all were in-service with 45 operational radars, some covering multiple airports in major metropolitan locations, across the United States & Puerto Rico. Several similar weather radars have also been sold to other countries such as China (Hong Kong). Funded by the United States Federal Aviation Administration (FAA), TDWR technology was developed in the early 1990s at Lincoln Laboratory, part of the Massachusetts Institute of Technology, to assist air traffic controllers by providing real-time wind shear detection and high-resolution precipitation data.

The primary advantage of TDWRs over previous weather radars is that it has a finer range resolution—meaning it can see smaller areas of the atmosphere. The reason for the resolution is that the TDWR has a narrower beam than traditional radar systems, and that it uses a set of algorithms to reduce ground clutter.

Characteristics

TDWR uses the frequency between 5600-5650 MHz (5 cm wavelength) carrier wave, with an angular resolution beam of 0.55 degrees and has a peak power of 250 kW. In reflectivity, the resolution in distance is 150 metres (500 ft) within 135 kilometres (84 mi) of the radar and 300 metres (1,000 ft) from 135 kilometres (84 mi) to 460 kilometres (290 mi) to the radar. The reason for this difference is that the width resolution being angular, at larger range the width of the beam becomes quite large and to obtain a better averaging of data in a resolution volume, one has to increase the number of range pulse bins. This cut off is arbitrarily set for the software at 135 kilometres (84 mi).

In radial velocities, data are available up to 90 kilometres (56 mi) from the radar with the full angular resolution of 0.5 degrees and range resolution of 150 metres (490 ft). Because of the Pulse Repetition Frequency (PRF) used, there is aliasing and the maximum non-ambiguous velocity is 20 to 30 knots (23 to 35 mph; 37 to 56 km/h).

TDWR can perform near-surface scans at a 0.1-0.3 degree angle of inclination from the Earth's surface every minute. It can also perform composite scans in which the radar observes at several different angles of inclination in order to obtain a fuller picture of the atmospheric conditions; each such composite scan requires 6 minutes.

Comparison with NEXRAD

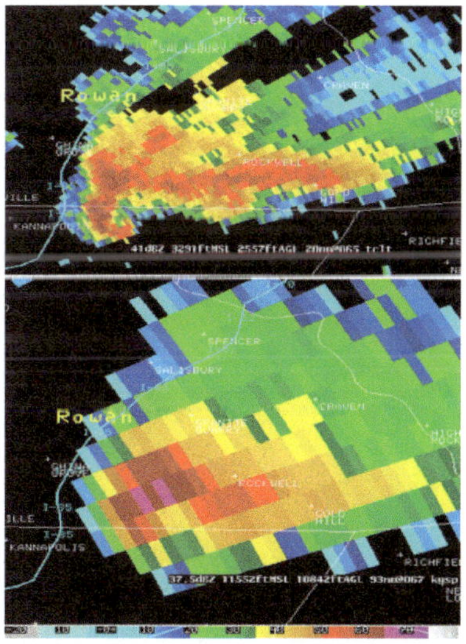

A TDWR return (top) and NEXRAD return (bottom) showing the improved resolution in reflectivity, but also showing the attenuation in the TDWR due to absorption from heavy precipitation as a black gap.

Advantages

A NEXRAD weather radar currently used by the National Weather Service (NWS) is a 10 cm wavelength (2700-3000 MHz) radar capable of a complete scan every 4.5 to 10 minutes, depending on the number of angles scanned, and depending on whether or not MESO-SAILS is active, which adds a supplemental low-level scan while completing a volume scan. Its resolution is 0.5 degrees in width and 250 metres (820 ft) in range. The non-ambiguous radial velocity is 62 knots (71 mph; 115 km/h) up to 230 kilometres (140 mi) from the radar.

The range resolution of the TDWR is nearly twice that of the NEXRAD. This will give much better details on small features in precipitation patterns, particularly in thunderstorms, in reflectivity and radial velocity. However, this finer resolution is only available up to 135 kilometres (84 mi) from the radar; beyond that, the resolution is close to that of the NEXRAD.

Shortcomings

The TDWRs and NEXRADs complement each other with overlapping coverage, each designed to optimally view different airspace regimes. TDWR's rapid update rate over short range (55 nmi range) captures microscale weather events quickly in terminal airspace. NEXRAD is a long range radar (200 nmi range) designed to serve multiple en route functions at high altitude, above terminal airspace, and far between terminals. NEXRAD's slower update rate covering a wider volume, captures mesoscale weather events. The shorter 5 centimetres (2.0 in) wavelength, which is closer to the size of a raindrop than the 10 centimetres (3.9 in) wavelength, is partially absorbed by precipitation. This is a serious drawback to using TDWR, as the signal can be strongly attenuated in heavy precipitation. This attenuation means that the radar cannot "see" very far through heavy rain and could miss severe weather such as strong thunderstorms which may contain the signature of a tornado, when there is heavy rain falling between the radar and that storm. When heavy rain is falling on the radome, the range of the TDWR is further limited. Finally, hail in a thunderstorm scanned by a TDWR can entirely block the signal as its size is larger than the wavelength.

A second problem is the smaller non-ambiguous radial velocity or Nyquist velocity. In the case of the TDWR, this means the velocity of precipitations moving at a speed beyond 30 knots (35 mph; 56 km/h) away or toward the radar will be analyzed incorrectly because of aliasing. Algorithms to correct for this do not always yield the proper results. NEXRAD has a threshold that is twice as high (62 knots (71 mph; 115 km/h)) and thus less processing and interpretation are needed. Because of this, the resolution of radar reflectivity for small scale features such as mesocyclones might be better in TDWR, but the velocity resolution may be worse, or at the very least incorrectly analyzed.

Thus, it is best to use the TDWR in conjunction with a traditional NEXRAD nearby to ensure that nothing is missed. In contrast to NEXRAD, which has national coverage of the contiguous United States (although with some holes due to terrain), TDWR has sporadic coverage meant for major airports. While certain areas of the country (the Northeast megalopolis, the states of Ohio and Florida, and the southwestern quarter of Tornado Alley in Oklahoma and Texas) have a high density of TDWR units, others (the entire West Coast, the northern Great Plains and Rocky Mountains, portions of the Deep South, and a stretch running from northern Pennsylvania through upstate New York and into northern New England) have no TDWR coverage at all.

Data Processing Improvements

The National Severe Storms Laboratory (NSSL) has a program of development and improvement of radar products extracted from data obtained from TDWR and NEXRAD radars. The *Severe Weather Warning Applications and Technology Transfer* (SWAT) group is sponsored by the National Weather Service and the FAA. It is working in 2009 on better filtering of non-weather echoes, better dealiasing algorithms of velocities, techniques to extract the horizontal component of the wind field from one or multiple radars. NSSL has been providing TDWR data to NWS office since the late 1990s. The NWS's Radar Operations Center (ROC), although focused on the NEXRAD network, also works with TDWRs.

Millimeter Cloud Radar

MIRA36 cloud radar at Tito Scalo (Potenza, Italy).

Millimeter-wave cloud radars, also denominated cloud radars, are radar systems designed to monitor clouds with operating frequencies between 24 and 110 GHz (Table below). Accordingly, their wavelengths range from 1 mm to 1.11 cm, about ten times shorter than those used in conventional S band radars such as NEXRAD.

ARM Millimeter Cloud Radar (MMCR) operating at 35 GHz.

Purpose

Table: Millimetre Cloud Radar Operating Bands.

Band name	Frequency range [GHz]	Wavelength range [mm]
Ka	24-40	7.5-11.1
W	75-100	2.7-4.0

The main purpose of these radars is the investigation of cloud properties and evolution. They typically operate at 35 GHz in the Ka band and at 94 GHz in the W band, where the atmospheric transmission is maximum. These frequencies correspond to wavelengths around 8 and 3 mm. Millimeter cloud radars have a high temporal and range resolution: temporal resolution is adjustable and typically ranges from 1 to 10 seconds, while the range resolution ranges from a few meters for cloud radars using frequency modulation of the transmitted signal (e.g. 4 m for the Frequency-Modulated Continuous-Wave (FMCW) 94 GHz cloud radar operating at UK Met Office), to several meters for magnetron based cloud radars (e.g. from 15 to 60 m for MIRA systems). The maximum detection range is between 14 and 20 km and the Doppler velocity resolution is of few cm/s. Most of the cloud radars are polarimetric, being able to measure the particles irregularity by means of the linear depolarization ratio (LDR). Usually they operate pointing to the zenith, but an increasing number of them have nowadays scanning units, that allow to retrieve additional information such as volume information when performing various Range Height Indicator (RHI) scans at different angles and at a relatively high speed, and vertical wind profiles, when performing Plan Position Indicator (PPI) at few degrees from the zenith.

Regarding the different cloud radars operating wavelengths, it must be taken into account that longer wavelengths are less attenuated by drizzle and rain, while shorter wavelengths are more sensitive to smaller particles. Accordingly, radar signals are less attenuated in Ka-band than in W-band, while W-band radars are able to detect smaller particles. Even though, the cloud radar beams are much less attenuated by cloud or precipitation particles than the lidar beam.

Cloud radars are used nowadays to determine cloud boundaries (e.g. cloud bases and tops) and to estimate clouds microphysical properties, such as particle size and mass content, which aids in understanding how clouds reflect, absorb and transform radiant energy passing though the atmosphere. They are also intensively used to investigate fog. Furthermore, cloud radars are being used for entomology studies for more than 40 years, since Ka and W-band radars detect almost exclusively insect targets on warm cloudless days, and more recently are also used for the study of giant aerosols.

Even though most cloud radar systems are ground-based, they can also be airborne and space-based. Examples of airborne systems are cloud radars installed on HALO (High Altitude and Long Range Research Aircraft) and Wyoming KingAir Research Aircraft. An example of spaceborne Cloud Profiling Radar (CPR) is operating in the CloudSAT satellite since 2006. The first spaceborne CPR with Doppler capability will be launched in 2016 within the Earth Clouds, Aerosols and Radiation Explorer (EarthCARE) mission.

Measuring with a Radar: from IQ to Spectra

Pulsed Radar systems are considered active instruments because they transmit in the atmosphere an electromagnetic wave and receive the signal reflected back from the atmosphere. In such radars, the antenna sends out the electromagnetic wave, and then collects the returned signal. Radars are composed of different hardware parts, each of which contains different elements. The figure on page 9 in Clothiaux et al. 1996 is showing such units.

The electromagnetic wave sent in the atmosphere is an EM wave of the form shown in figure on page 10 of. Such wave is generated by an oscillator in the transmitter unit and then transferred through wave guides to the antenna, which radiates it in the atmosphere. The theory of propagating waves in rectangular shaped waveguides having a vertical axis of symmetry shows that by setting appropriate dimensions of the waveguide, the resulting electric field is propagating parallel to the interior space of the waveguide in the vertical direction, having a y sinusoidal in time component (transverse wave).

The expression of the electric field radiated at a distance r far from the antenna is, adopting complex notation in spherical coordinate system $(r;\theta;\phi)$:

$$E_i(r,\theta,\phi,t) = \frac{A_i(\theta,\phi)e^{i\phi t}}{r} e^{i(k_c r - \omega_c t)} U(t - 2r/c) \quad (1)$$

where ω_c is the carrier frequency of the wave, t is the time, c is the speed of light, r is the distance from the antenna, k_c is the wavenumber and λ_c is the wavelength, while A_i is the amplitude of the wave which depends on the power delivered to the antenna, on its characteristics and it is affected by the power losses in the waveguides. The function $U(t-2r/c)$ is a modulating function which is 1 when its argument is between 0 and τ, and 0 elsewhere. Therefore, such electromagnetic (EM) field is oscillating sinusoidally in time within the pulse width T_{pw} and is zero outside the pulse envelope. This EM wave is sent to the atmosphere: every pulse is scattered by a volume of air filled with hydrometeors and comes back to the radar. The antenna collects the returned signal, which is then filtered to remove the high carrier frequency, amplified and then downconverted and digitized.

The scattered electric field collected by the antenna is a composite of discrete echoes from all the scatterers contained in the volume and can be written as:

$$E_r(t) = \sum_{m=0}^{N_s} A_{r,m} e^{i(2k_c r_m - (\omega_c - \omega_{d,m})t + \phi_{s,m} + \phi_t)} U(t - r/c) \quad (2)$$

where $A_{r,m}$ is the amplitude of the electric field scattered by the mth scatterer, r_m is the position of the mth scatterer, ω_c is the carrier frequency and $\omega_{d,m}t$ represents the shift in phase of the scattered wave in the direction of the radar due to the relative radial velocity of the target with respect to the radar, while $\phi_{s,m}$ and ϕ_t are respectively the phase shift upon scattering and the transmitter phase, which can be assumed constants ($\phi_{s,m}$ can depend on time for meteorological targets like vibrating water drops and tumbling ice particles).

As already mentioned, a weather signal is a composite of echoes coming from a very large number of hydrometeors. Such echoes are received continuously at the radar antenna after a delay equal to the time taken by the wave to reach the target and come back to the radar. Since individual echoes are impossible to be resolved individually, we sample the signal coming from the atmosphere at given discrete time delays τ_s.

Such τ_s defines the range of the scatterers that contribute mostly to the sample of the signal. The number of distinct resolution volumes in space from which the radar collects information is equal to the number of samples N_g that are collected by the radar between any two radar pulses. The expression of the sampled received voltage is:

$$V(\tau_s) = \sum_{m=0}^{N_g} A_{r,m} e^{i(2k_c r_m - (\omega_{d,m})t + \phi_{s,m})} \quad (3)$$

For every range gate (τ_s) and for every pulse cycle (T_s), the so-called I and Q voltages are derived taking the real and imaginary part of the complex voltage given in (3): Their expressions are:

$$I(\tau_s, T_s) = Re[\sum_{i=0}^{N_g} A_{r,m} e^{i(2k_c r_m - (\omega_{d,m})t + \phi_{s,m})}] \quad (4)$$

$$Q(\tau_s, T_s) = Im[\sum_{i=0}^{N_g} A_{r,m} e^{i(2k_c r_m - (\omega_{d,m})t + \phi_{s,m})}] \quad (5)$$

Therefore, after such sampling of the received signal, a time serie of I/Q signals is associated to every range gate.

As radar transmits coherent pulses with a certain pulse repetition frequency (PRF), the doppler processing technique is applied. A Power spectrum can be calculated from a sequence of $N_f I_n + iQ_n$ components in the following way. In every range gate the FFT algorithm is applied to the serie of N_f IQ signals, where N_f is the number of FFT points used in the FFT. The result of the FFT is a complex spectrum $S_{compl}(k)$ where k is the Doppler frequency. The power spectra is then easily estimated by

$$S(k) = S_{compl} * S^*_{compl}(k) \quad (6)$$

Such spectra is the distribution of radial velocities of the targets in the volume analyzed weighted with the power of the signal.

The Doppler Spectra

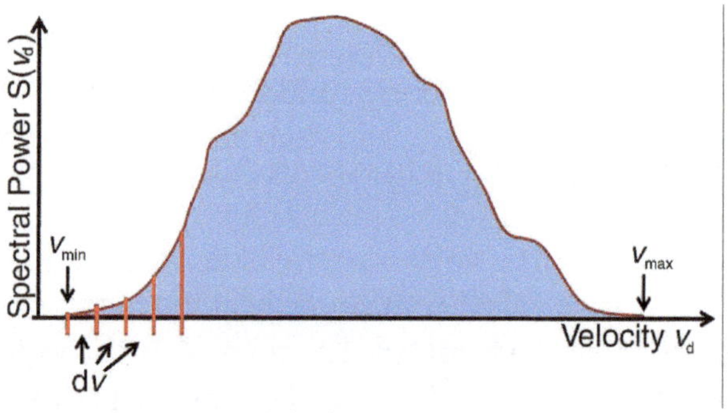

Sketch of a Doppler spectrum. The spectral power $S(v_d)$ is represented as a function of the Doppler velocity v_d. The parameters v_{min} and v_{max} denote the minimum and the maximum velocities, at which $S(v_d) > 0$, and dv is the velocity resolution.

Usually numerous targets are present in the sample volume of the radar. Each of the individual targets produces a frequency shift according to its radial velocity. Measuring the returned power in any interval of the frequency shift allows the detection of the Doppler spectrum. This is illustrated in the Figure on the right, where a sketch of the Doppler spectrum is shown. The frequency is already converted into velocity in the x axis.

From the Doppler spectrum the reflectivity Z can be calculated using the expression:

$$Z = \int_{v_{min}}^{v_{max}} S(v_d) dv_d \qquad (7)$$

The integral of the spectral power $S(v_d)$ is called the 0-moment of the Doppler spectrum, which equals the reflectivity Z. Further, the first moment (8) and the second moment (9) of the Doppler spectrum can be calculated:

$$V = \frac{1}{Z} \int_{v_{min}}^{v_{max}} v_d S(v_d) dv_d \qquad (8)$$

$$W = \sqrt{\frac{1}{Z} \int_{v_{min}}^{v_{max}} (v_d - V)^2 S(v_d) dv_d} \qquad (9)$$

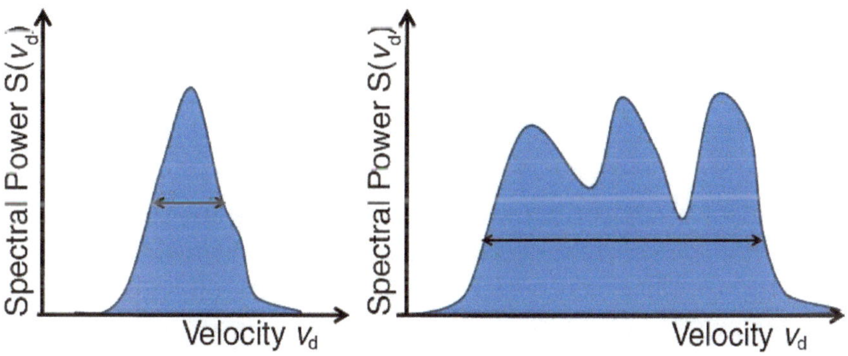

Sketch of the Doppler width. It shows the relation between the Doppler width W, depicted by the arrows, and the number of modes of the droplet size distribution.

The first moment of the Doppler spectrum is the mean Doppler velocity V. It corresponds to the mean radial velocity in the sampled volume.

The second moment is called the Doppler width W. It corresponds to the variance of the Doppler spectrum. The Doppler width basically is a measure of the width of the spectrum of the detected velocities. It can also be a measure of the number of modes the targets size distribution has. A mono-modal distribution gives a small spectral width. A distribution that has several modes yields a lager spectral width, as illustrated in Figure 5. The spectral width is also influenced by turbulent motion of the targets in the sample volume: the spectral width increases with increasing turbulence.

The expressions to calculate the Kurtosis and the Skewness are coming out of mathematical statistics several formulations are used. Some of them can be found in the quoted literature in Kurtosis section.

Doppler Velocity

Mean Doppler velocity can be seen as the mean frequency of a power spectrum (Doppler velocity) of the backscattered power.

Given the mean Doppler velocity (mean frequency of the power spectrum) of the measured volume along the line of sight - the radial velocity. For vertical measurements, the Doppler velocities is the sum of terminal velocity of particles, caused by gravity force, and air motions within the measured volume. Terminal velocity contains the information about microphysical properties of cloud particles. In order to obtain the terminal velocity from the cloud-radar observations, the influence of air motions that shift the spectrum have to be removed from the Doppler spectra, improving in this way the microphysics representativity. An approach to correct this shift is given by Kollias et al., improving the accuracy of the relation between the corrected Doppler velocity and the particle fall velocities.

Doppler Width

The Doppler width, or Doppler spectrum width, is the standard deviation of the spectrum. Accordingly, small values stand for narrow spectra, while higher spectrum width corresponds to a bigger spreading of the hydrometeors over the velocity domain (frequency domain). Reasons for a higher width could be bi- or multiple-modal spectra. This means that the measured volume contains more than one hydrometeor population, which leads to two or more modes in the Doppler spectrum because of the different terminal velocities. A rapid change in Doppler width in combination with the mean Doppler velocity can also be used to separate two hydrometeor population.

Since the air motion in the atmosphere affects the Doppler width, this parameter provides information about turbulence within the measured volume. Up- and downdrafts slow down the terminal fall velocities and can decrease or increase the Doppler width. If the measurements are not done pointing to the zenith, the horizontal wind component influences the width as well. The knowledge of the actual wind components in the atmosphere can be used to correct the Doppler spectrum, so that the retrieved microphysical parameters can be improved and the uncertainties decreased.

Skewness

The skewness parameter of the Doppler spectrum describes the asymmetry of the spectrum with respect to a symmetric Gaussian distribution.

$$Sk = \frac{\frac{1}{Z}\int_{v_{min}}^{v_{max}} (v_d - V)^3 \,|\, S(v_d) \,|^2 \, dv_d}{W^3} \qquad (10)$$

This parameter relates the location of the spectrum peak in respect to the mean spectrum value. Therefore, a positive skewness value indicates that the peak is located to the left with respect to the mean. A negative skewed spectrum has its peak on the right side with respect to the spectrum mean. A value around zero indicates a symmetric spectrum. In this way, the shape of a Doppler gives information on changes of cloud microphysics or of dynamical changes within the measured volume. The higher the elevation of the radar is, the higher are the dynamical influences on this parameter. Because wind shears lead to a broadening of the Doppler spectrum width, they can lead to rapid changes in skewness as well. To be able to relate the change in skewness of the Doppler spectrum, the Doppler width should be cross-checked as well.

If the radar measurements are done vertically, the skewness of the Doppler spectrum gives information about measured cloud microphysics. The horizontal wind within the measured volumes cause only a shift of the whole spectrum within the Doppler domain. This means that the spectrum is shifted along the Doppler velocity axis but it did not influence the broadening of the spectra. So the changes in skewness give information about:

- if the minority of the hydrometeors is falling faster or slower as the mean Doppler velocity
- if the Doppler spectrum is dominated by small hydrometeors (slower than the mean) or bigger particles (faster than the mean Doppler velocity)
- the shape of the size distribution of the measured hydrometeors
- changes over height or time can be related to a change in the cloud microphysics

Kurtosis

The kurtosis of the Doppler spectrum is an asymmetry parameter too. It describes the steepness of the spectrum curve.

$$K = \frac{\frac{1}{Z}\int_{v_{min}}^{v_{max}} (v_d - V)^4 \, |S(v_d)|^2 \, dv_d}{W^4} \tag{11}$$

So it gives the peakedness (width of peak) of the given spectrum. Therefore, kurtosis can give information about the spectrum tail weight or lack of shoulders, helping to describe the spectrum shape better.

If a Doppler spectrum would be exactly normal distributed its kurtosis would be 3. If in general the kurtosis is positive the spectrum is called leptokurtic, or leptokurtotic. Most known distribution shapes would be those that follow a normal distributed spectra shape. It can be assumed that the Doppler spectrum is dominated by one single particle population that leads to a strong and narrow peak. If the spectrum has a negative kurtosis it is called platykurtic, or platykurtotic. The shape of such a spectrum has a has lower, wider peak around the mean and thinner tails.

The Doppler radar kurtosis analysis started quite recently, so there are still few scientific publications dealing with this parameter. An example can be find in Kollias et al., where kurtosis is used to interpret Doppler spectra and to understand the microphysical changes that it represents.

Polarimetric Measurements in Cloud Radars

Polarimetric methods are considered to be a powerful tool in the remote sensing of the atmosphere, and particularly in radar observations of clouds and precipitation. Polarimetric techniques have been well developed for operational use in weather radar networks such as American NEXRAD and European OPERA, and, currently, are being implemented in ground based and airborne cloud radars. These techniques enable advanced clutter filtering, discrimination between meteorological and non-meteorological targets, and classification of atmospheric scatterers. Another potential of polarimetric measurements of cloud radars is the estimation of ice crystals habit that is one of the main problems in remote observations of mixed-phase clouds. The assumed shape of ice crystals is considered to be the main source of errors in size and number concentration retrievals based on vertical combined lidar and radar observations. Furthermore, in the existing numerical weather prediction models the assumed shape defines the depositional growth rate of ice particles and the area-mass-terminal velocity relations of ice crystals. Thus, ice habit can lead to significant uncertainties.

A number of different polarimetric configurations were evaluated in theoretical and experimental studies and are explained below.

Most of the existing pulse cloud radars operate in the LDR-mode. In this mode, a radar transmits a horizontally polarized radiation and receives horizontal and vertical components of the scattered radiation in co-polarized and cross-polarized channels, respectively. The ratio of the power in the cross-polarized channel over the power in the co-polarized channel, named as linear depolarization ratio, is the polarimetric variable obtained by cloud radars of this type. Often cloud radars have two receiving units, which allows simultaneous measurements of the orthogonal components of the received signal. Some cloud radars have only one receiving unit and measure the orthogonal components consequently, with pulse-to-pulse switching of the receiving channel. Many of operational cloud radars with LDR-mode are pointed vertically together with other remote sensing instruments in order to retrieve microphysical properties of cloud particles. LDR, measured by vertically pointed cloud radar, is used for the detection of the melting layer (also denominated as bright band) and clutter filtering. The application of LDR-mode for the shape classification in scanning cloud radars is limited by its high sensitivity to the orientation of scatterers.

Some cloud radars operate in the SLDR-mode, that is a modification of traditional LDR-mode. In this mode the antenna system of the radar is rotated by 45°, i.e. the transmitted signal has +45° or -45° polarization. In contrast to LDR-mode, SLDR-mode is less sensitive to the particles orientation and, therefore, it is used in cloud radars for the classification of ice crystals. It should be noted that determination of ice crystals' habit by ground based cloud radars requires elevation scanning and is based on analysis of the angular changes of the polarimetric variables. Evaluation of SLDR with in-situ observations showed the possibility to discriminate between rounded graupel, dendrite crystals, and aggregated ice particles.

Polarization-agile cloud radars use pulse-to-pulse switching between horizontal and vertical polarization state of the transmitted wave. In addition to LDR, such systems are able to measure differential reflectivity (Z_{DR}) and correlation coefficient (ρ_{HV}). The combined analysis of LDR, Z_{DR}, and ρ_{HV} can be used not only for classification of ice particle habit but also for characterization of their orientations.

CDR-mode cloud radars transmit a circularly polarized wave and receive co-polarized and cross-polarized components. The output polarimetric variable is the so-called circular depolarization ratio (CDR), and it is calculated as the ratio of cross-polarized power over the co-polarized power. In comparison with LDR, CDR is not affected by particles aligned in polarization plane, e.g. insects or ice crystals in electrified atmospheric conditions.

Measurement Example

As the cloud radars sample the atmosphere, they measure the backscatter signal produced by different hydrometeor types (cloud droplets, drizzle, rain drops, ice particles, snow, etc.) and non-hydrometeorological targets. All these targets have different vertical velocities and shapes, and therefore the Doppler spectra and LDR can be used to distinguish between targets. In the figure below, an example of the MIRA-36 cloud radar measurements in Potenza is shown, where using the information given by the different parameters, an identification of targets can be done:

- Liquid water clouds produce a signature in the reflectivity but not in the LDR, since the liquid droplets are almost spherical.
- Ice clouds are characterized by relatively high LDR signals due to their irregular shapes.
- Insects produce high LDR values and are usually within the Convective Boundary Layer (CBL), where up- and downdrafts are observed continuously.
- The melting layer can be identified by an enhancement of the LDR, which is caused by irregular ice particles coated by liquid water during the phase transition.
- Rain is characterized by high reflectivity values, high falling velocities and a broadening of the spectra.
- Liquid layers within mixed-phase clouds can be detected by slightly higher peak width values, in accordance to the higher turbulence within.

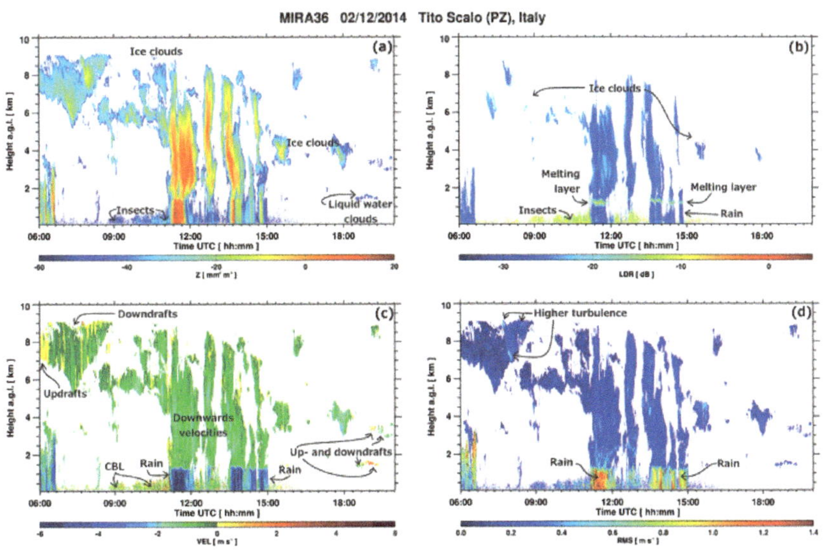

Measurements performed by a MIRA-36 system in Tito Scalo, Potenza (Italy) on 2nd December 2014: (a) equivalent reflectivity, (b) Linear Depolarization Ratio, (c) Doppler velocity, where negative values indicate movement towards the radar, and (d) peak width.

Cloud Radar Systems

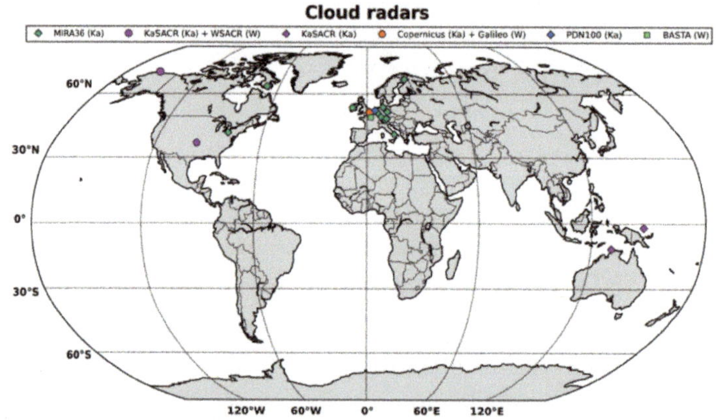

Cloud radar instruments involved in ACTRIS and NOAA projects.

The National Oceanic and Atmospheric Administration (NOAA) designed the Scanning ARM Cloud Radar (SACR) to monitor clouds overhead at various testing sites of the U.S. Department of Energy's atmospheric radiation measurement program. The radars operate continuously at these sites in Oklahoma, Alaska and the tropical western Pacific Ocean, and are designed to function for at least ten years with minimal manned attention. The SACR s operate at a frequency of 35 and 94 GHz, being designated as KaSACR and WSACR respectively.

Since the late 2000s, a meteorological 35.5 GHz Ka-band commercial cloud radar (MIRA-36) designed by METEK GmbH in collaboration with the Institute for Radio Astronomy, Kharkov (Ukraine) is in the market. Nowadays, there are eleven systems monitoring the cloud properties continuously in different sites, most of them in Europe. Cloudnet, a network of stations for the continuous evaluation of cloud and aerosol profiles in operational NWP models, assimilates their data, as well as data acquired by other instruments in the various sites with this purpose.

Outside of these networks, some research sites are also operating cloud radars. For example, at the Chilbolton Observatory (UK), two mm-wavelength radars are operating continuously at 3.2 and 9 mm for the study of insects migrations, while a 35 GHz cloud radar is operating in Cabauw (the Netherlands).

References

- "$400m Hornet radar contract 'dumped' – Breaking News – National – Breaking News". The Age. Australia. 13 September 2006. Retrieved 15 February 2012
- Helgason, Guðmundur. "Discussion Forums :: Technology and Operations :: RE: U-boat Radar". German U-boats of WWII - uboat.net. Retrieved 3 June 2010
- Brown, Louis (1999). A Radar History of World War II: Technical and Military Imperatives. London: Institute of Physics Publishing. pp. 315–316. ISBN 0-7503-0659-9
- "Development of Airborne Early Warning and Control System" (Press release) Press Information Bureau, GoI. 11 December 2003. Retrieved 2008-07-25
- Staff. Campaign Diary December 1943, Royal Air Force Bomber Command 60th Anniversary web site. Retrieved 11 August 2008

4

Technological Aspects of Radar Systems

The technological aspects discussed in this section are radar altimeter, radar detector, terrain-following radar, precision approach radar and radar warning receiver. Radar altimeter is used on aircrafts to measure the terrain beneath the aircraft whereas a radar detector is used in motorcycles to detect its speed. The aspects elucidated in this chapter are of vital importance, and provide a better understanding of radar systems.

Radar Altimeter

A radar altimeter, electronic altimeter, reflection altimeter, radio altimeter (RADALT), low range radio altimeter (LRRA) or simply RA, used on aircraft, measures altitude above the terrain presently beneath an aircraft or spacecraft by timing how long it takes a beam of radio waves to reflect from the ground and return to the plane. This type of altimeter provides the distance between the antenna and the ground directly below it, in contrast to a barometric altimeter which provides the distance above a defined datum, usually mean sea level.

ITU Definition

From the legal point of view, a radio altimeter is – according to *article 1.108* of the International Telecommunication Union´s (ITU) ITU Radio Regulations (RR) – defined as "*Radionavigation equipment, on board an aircraft or spacecraft, used to determine the height of the aircraft or the spacecraft above the Earth's surface or another surface.*" *Radionavigation equipment* shall be classified by the *radiocommunication service* in which it operates permanently or temporarily. The utilization of radio altimeter equipment is categorised as so-called safety-of-life service, must be protected for Interferences, and is an essential part of Navigation.

Principle

As the name implies, radar (radio detection and ranging) is the underpinning principle of the system. The system transmits radio waves down to the ground and measures the time it takes them to be reflected back up to the aircraft. The altitude above the ground is calculated from the radio waves' travel time and the speed of light.

Radar altimeters normally work in the E band, K_a band, or, for more advanced sea-level measurement, S band. Radar altimeters also provide a reliable and accurate method of measuring height above water, when flying long sea-tracks. These are critical for use when operating to and from oil rigs.

Pulse-limited Altimetry

Consider a radar pulse emanating from a radar beacon traveling downwards and interacting with a flat ocean surface. The following figure shows an illustration of the vertical cross-section and top-down view of the radar pulse (adapted from).

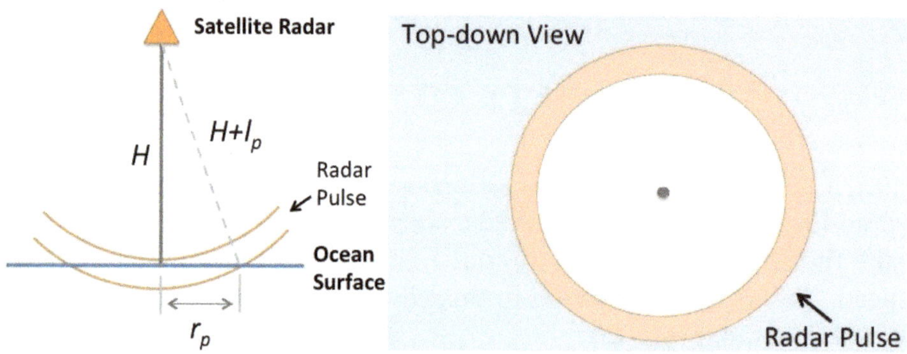

The radar altimeter measures the return power of the radar pulse that's reflected off the land/ ocean surface. The temporal evolution of the reflected radar pulse is interpreted in order to estimate the distance between the radar altimeter and the reflecting surface; surface irregularities can also be estimated. The expected return pulse can be derived from a few basic mathematical considerations.

From the previous illustration of the radar pulse vertical cross-section, the radius of the outer edge can be found using the Pythagorean theorem:

$$H^2 + r_p^2 = (H + l_p)^2 = H^2 + l_p^2 + 2Hl_p$$

where r_p is the leading edge of the pulse. If we assume l_p^2 is small and can be neglected then we can solve for r_p as:

$$r_p = (2Hl_p)^{\frac{1}{2}} = (Hct_p)^{\frac{1}{2}}$$

The time evolution of the return power measured from the footprint of the radar signal reflecting off the ocean/land surface can be described in three parts: (1) the time before the pulse arrives, (2) the time after the beginning of the pulses arrival and before of the tail of the pulse has arrived, and (3) after the tail of the pulse has arrived. The power function can then be expressed functionally as:

$$P(t) = \begin{cases} 0 & t < t_o \\ \pi r^2(t) & t_o < t < t_o + t_p \\ \pi[r^2(t) - r^2(t - t_p)] & t > t_o + t_p \end{cases}$$

Using the equation for r_p and normalizing by the peak power yields:

$$P(t) = \begin{cases} 0 & t < t_o \\ \dfrac{t - t_o}{t_p} & t_o < t < t_o + t_p \\ 1 & t > t_o + t_p \end{cases}$$

This following plot demonstrates the power function for a radar wave pulse. The power function should be constant with time after the full pulse reaches the ocean/land surface (at $(t-t_o)/t_p = 1$), however the power will actually decrease with time w.r.t the illumination pattern of the radar on the ocean/land surface.

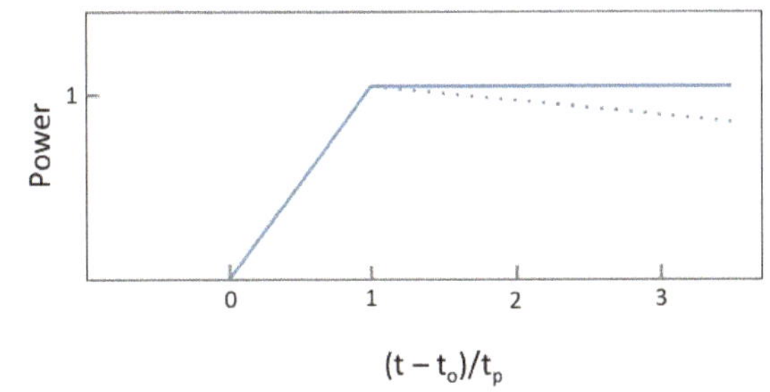

Frequency-modulated Continuous-wave Radar Altimetry

Alternatively, Frequency Modulated Continuous-wave radar can be used. The greater the frequency shift the further the distance traveled. This method can achieve much better accuracy than the pulsed radar for the same outlay and radar altimeters that use frequency modulation are industry standard.

As of 2010, all commercial radar altimeters use linear frequency modulation - continuous wave (LFM-CW or FM-CW). As of 2010, about 25,000 aircraft in the US have at least one radio altimeter. That includes all commercial transport and all business aircraft licensed to fly for hire, which are required to have at least 2 radio altimeters per aircraft.

Delay-doppler (or SAR) Altimetry

The primary difference between delay-doppler (or Synthetic Aperture Radar) and pulse-limited altimetry is that delay-doppler altimetry looks at a smaller section of the pulse-limited radar footprint, but emits far more pulse signals to give the effect of covering the same footprint as pulse-limited but with better resolution. The top-down view in the figure below shows the decreased footprint size of the delay-doppler signal.

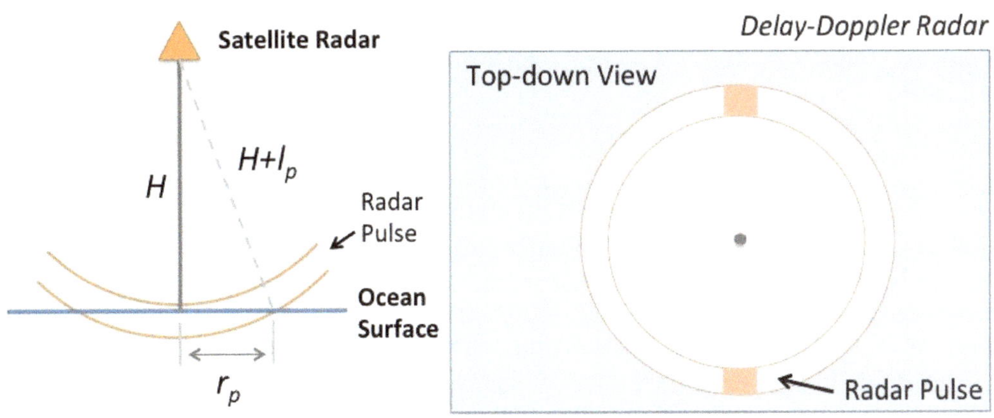

In order to determine the power signal of the delay-doppler radar as a function of time, we'll need to assume that the footprint of the pulsed radar is small enough to be considered two rectangles of width W. This allows the power function to be similarly written as:

$$P(t) = \begin{cases} 0 & t < t_o \\ 2Wr(t) & t_o < t < t_o + t_p \\ 2W[r(t) - r(t - t_p)] & t > t_o + t_p \end{cases}$$

Using the equation for the radius derived in the previous section, the power vs. time function, normalized by the time relative to the arrival time at the leading edge, $t' = t - t_o$, becomes:

$$P(t') = \begin{cases} 0 & t' < 0 \\ (\frac{t'}{t_p})^{\frac{1}{2}} & 0 < t' < t_p \\ (\frac{t'}{t_p})^{\frac{1}{2}} - (\frac{t'}{t' - t_p})^{\frac{1}{2}} & t' > t_p \end{cases}$$

An illustration of the normalized power vs. time function is shown below:

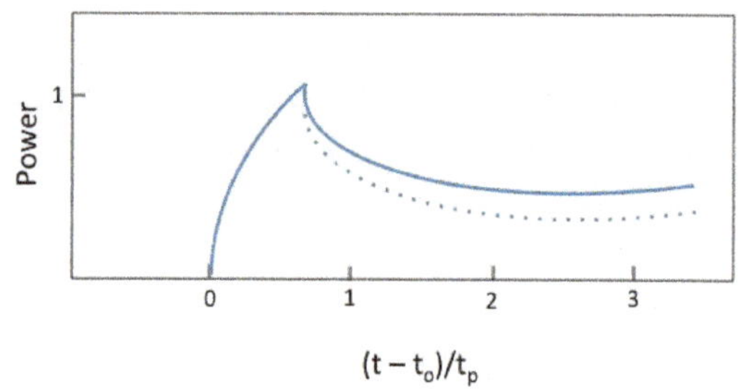

There are two primary advantages that the delay-doppler method has over the traditional pulse-limited altimeter. First, the radar footprint is covering less area so the radar pulse emitted by the satellite requires much less power. It covers the radar footprint using more frequent, but less power consuming radar pulses. Secondly, the return waveform has a more complex signature allowing the arrival time of the pulse to be more accurately constrained.

Error Budget

The following list notes the error sources associated with radar altimetry measurements:

Tides - Tidal variations are much larger than the dynamic variations in sea surface height. Because tidal periods can be on the order of diurnal and semidiurnal, the tides create an aliased frequency in the temporal variations in the sea level height that must be removed.

Electromagnetic Bias - There is a sea state bias where the troughs of waves tend to focus waves back to the radar, while the crests of the waves scatter waves away from the radar.

Ionosphere - The ionosphere can also impose a delay on the radar return signal, where electron plasma in the ionosphere slow down the group velocity of the radar pulse. The electron density in the ionosphere varies throughout the day, complicating the ionosphere correction.

Dry Troposphere - Refraction from the dry gas component of the atmosphere create a signal delay in the radar, but a correction can be approximated using the Saastamoinen formula: ΔRdry= -0.02277Po*(1+0.0026cos2φ) where Po is the sea level pressure in Pascal and φ the latitude.

Wet Troposphere - Water vapor can also cause a delay in the radar signal which can be more difficult to correct. A delay correction for the total water column in the radar measurement can be accounted for using output from meteorological models, like ECMWF and NCEP.

Invention

In 1924, American engineer Lloyd Espenschied invented the radio altimeter. In 1938, Bell Labs put Espenschied's device in a form that was adaptable for aircraft use. In 1938 in co-operation with Bell Labs, United Air Lines fitted a radar type device to some of its airliners as a terrain avoidance device.

Civil Aviation Applications

Radar altimeters are frequently used by commercial aircraft for approach and landing, especially in low-visibility conditions and automatic landings, allowing the autopilot to know when to begin the flare maneuver. Radar altimeters give data to the autothrottle which is a part of the Flight Computer.

Radar altimeters generally only give readings up to 2,500 feet (760 m) above ground level (AGL). Frequently, the weather radar can be directed downwards to give a reading from a longer range, up to 60,000 feet (18,000 m) above ground level (AGL). As of 2012, all airliners are equipped with at least two and possibly more radar altimeters, as they are essential to autoland capabilities. (As of 2012, determining height through other methods such as GPS is not permitted by regulations.) Older airliners from the 1960s (such as the British Aircraft Corporation BAC 1-11) and smaller airliners in the sub-50 seat class (such as the ATR 42 and BAe Jetstream series) are equipped with them.

Radar altimeters are an essential part in ground proximity warning systems (GPWS), warning the pilot if the aircraft is flying too low or descending too quickly. However, radar altimeters cannot see terrain directly ahead of the aircraft, only that below it; such functionality requires either knowledge of position and the terrain at that position or a forward looking terrain radar. Radar altimeter antennas have a fairly large main lobe of about 80° so that at bank angles up to about 40°, the radar detects the range from the aircraft to the ground (specifically to the nearest large reflecting object). This is because range is calculated based on the first signal return from each sampling period. It does not detect slant range until beyond about 40° of bank or pitch. This is not an issue for landing as pitch and roll do not normally exceed 20°.

The altitude specified by the device is not the indicated altitude of the standard barometric altimeter. A radar altimeter measures absolute altitude - the height Above Ground Level (AGL). Absolute altitude is sometimes referred to as height because it is the height above the underlying terrain.

Military Aviation Applications

Radar altimeters are also used in military aircraft to fly quite low over the land and the sea to avoid radar detection and targeting by anti-aircraft guns or surface-to-air missiles. A related use of radar altimeter technology is terrain-following radar, which allows fighter bombers to fly at very low altitudes.

The F-111s of the Royal Australian Air Force and the U.S. Air Force have a forward-looking, terrain-following radar (TFR) system connected via digital computer to their automatic pilots. Beneath the nose radome are two separate TFR antennae, each providing individual information to the dual-channel TFR system. In case of a failure in that system, the F-111 has a back-up radar altimeter system, also connected to the automatic pilot. Then, if the F-111 ever dips below the preset minimum altitude (for example, 15 meters) for any reason, its automatic pilot is commanded to put the F-111 into a 2G fly-up (a steep nose-up climb) to avoid crashing into terrain or water. Even in combat, the hazard of a collision is far greater than the danger of being detected by an enemy. Similar systems are used by F/A-18 Super Hornet aircraft operated by Australia and the United States.

The radio altimeter first showed up in the German Junkers Ju-87 "Stuka" dive bomber which was equipped with one for automatic pullouts in the dive bomb run which usually consisted of an 80-90 degree dive. The Stuka pilot would set the radio altimeter to 750m which was connected to the bomb release and automatic pull out. The radio altimeter would drop the bombs and pull the Stuka out of the dive at the set altitude to a level flight. This was invented because the pilots would black out on the pull out, usually for 2 to 5 seconds.

Radar Detector

An early radar detector

Car radar detector (Japanese)

A radar detector is an electronic device used by motorists to detect if their speed is being monitored by police or law enforcement using a radar gun. Most radar detectors are used so the driver can reduce the car's speed before being ticketed for speeding. Only doppler radar-based devices can be detected – other speed measuring devices including those using ANPR, piezo sensors, and VASCAR technology cannot be detected. LIDAR devices require a different type of sensor, although many modern detectors include LIDAR sensors. Most of today's radar detectors detect signals across a variety of wavelength bands: usually X, K, and K_a. In Europe the K_u band is common as well.

Description

One device law enforcement use to measure the expected speed of a moving vehicle is called Doppler radar, and it uses the Doppler effect to measure the relative speed of a vehicle. Namely, Doppler radar works by beaming a radio wave at the vehicle to then measure the expected change in frequency of the reflected wave (that bounces off the vehicle). Many times, law enforcement employ Doppler radar in their hand-held radar guns, on their vehicles, as well as on fixed objects such as traffic signals.

Radar detectors use a superheterodyne receiver to detect these electromagnetic emissions from the radar gun, and raise an alarm to notify the motorist when a transmission is detected. False alarms can occur however due to the large number of devices, such as automatic door openers (such as the ones at supermarkets) and adaptive automotive cruise control, that operate in the same part of the electromagnetic spectrum as radar guns.

In recent years some radar detectors have added GPS technology. This allows users to manually store the locations where police frequently monitor traffic, with the detector sounding an alarm when approaching that location in the future (this is accomplished by pushing a button and doesn't require coordinates to be entered). These detectors also allow users to manually store the coordinates of sites of frequent false alarms, which the GPS enabled detector will then ignore. Some GPS enabled detectors can download the GPS coordinates of speed monitoring cameras and red-light cameras from the Internet, which are contained in the TRINITY database. An interstate traveler could receive an alert when approaching a location with a speed monitoring camera.

Counter Technology

Radar guns and detectors have evolved alternately over time to counter each other's technology in a form of civilian electronic "warfare". For example, as new frequencies have been introduced, radar detectors have initially been "blind" to them until their technology, too, has been updated. Similarly, the length of time and strength of the transmissions have been lowered to reduce the chance of detection, which in turn has resulted in more sensitive receivers and more sophisticated software counter technology. Lastly, radar detectors may combine other technologies, such as GPS-based technology with a point of interest database of known speed trapping locations, into a single device to improve their chances of success.

Radar Detector Detectors

The superheterodyne receiver in radar detectors has a local oscillator that radiates slightly, so it is possible to build a radar-detector detector, which detects such emissions (usually the frequency of the radar type being detected, plus about 10 MHz). The VG-2 Interceptor was the first device developed for this purpose, but has since been eclipsed by the Spectre III and Spectre Elite. This form of "electronic warfare" cuts both ways - since detector-detectors use a similar superheterodyne receiver, many early "stealth" radar detectors were equipped with a radar-detector-detector-detector circuit, which shuts down the main radar receiver when the detector-detector's signal is sensed, thus preventing detection by such equipment. This technique borrows from ELINT surveillance countermeasures. In the early 1990s, BEL-Tronics, Inc. of Ontario, Canada (where radar detector use is prohibited in most provinces) found that the local oscillator frequency of the detector could be altered to be out of the range of the VG-2 Interceptor. This resulted in detector manufacturers responding by changing their local oscillator frequency. Today, practically every radar detector on the market is immune to the VG-2 Interceptor.. The VG-2 is no longer in production and radar detectors immune to the Spectre Elite are available.

Radar Scrambling

It is illegal in many countries to sell or possess any products that actively transmit radar signals intended to jam radar equipment. In the United States, actively transmitting on a frequency licensed by the Federal Communications Commission (FCC) without a licence is a violation of FCC regulations, which may be punishable by fines up to $10,000 and/or up to one year imprisonment.

Passport x50 Radar/Laser detector.

LIDAR Detection

Newer speed detection devices use pulsed laser light, commonly referred to as LIDAR, rather than radio waves. Radar detectors, which detect radio transmissions, are therefore unable to detect the infrared light emitted by LIDAR guns so a different type of device called a LIDAR detector is required. LIDAR detection, however, is not nearly as effective as radar detection because the output beam is very focused. While radar's radio waves can expand to 85 feet (26 m) across at 1,000 feet (300 m) from their source, LIDAR's light beam diffuses to only about 6 feet (1.8 m). Also, a police officer targeting a car will most likely aim for the centre mass or headlight of the vehicle and, because radar detectors are mounted on the windshield away from the beam's aim, they may not alert at all. Lastly, with such a focused beam, an officer using a LIDAR gun can target a single car in close proximity to others at ranges of up to 3,000 feet (910 m). This has resulted in some manufacturers producing LIDAR jammers. Unlike radar, LIDAR's frequencies and use are not controlled by the FCC. These jammers attempt to confuse police LIDAR into showing no speed on the display. Many times they are successful, and therefore many LIDAR manufacturers are producing LIDAR guns that have "jam codes" which show when they are being jammed. They will work on some LIDAR jammers and not others. In spite of this, police can often tell when they are being jammed when they get no reading on their LIDAR. Many jammer-equipped motorists using jammers now try to counter this by reducing their speed to legal limits before they turn their jammer equipment off or "kill the equipment" referred to as "JTK" or "Jam to Kill." Officers can often tell this is going on by just watching for their LIDAR equipment not being able to lock in a speed properly along with visual indication of sudden deceleration of the targeted vehicle. They will then pull the offending vehicle over and look for LIDAR jammers on the front of the vehicle and can hit the motorist with an obstruction of justice charge should they be found. Some states also have laws about any police radar or LIDAR jamming. These are CA, UT, CO, OK, TX, MN, IL, TN, VA, SC. In these states the penalties can be severe.

Despite the advent of LIDAR speed detection, radar remains more prevalent because of its lower price relative to LIDAR equipment and the amount of radar equipment already in service. In addition, in order to use LIDAR the officer must be stationary and cannot be moving at all on account of the need for a very precise aim.

Legality

Using or possessing a radar detector or jammer is illegal in certain countries, and it may result in fines, seizure of the device, or both. These prohibitions generally are introduced under the premise that a driver who uses a radar detector will pose a greater risk of accident than a driver who does not. The table below provides information about laws regarding radar detectors in particular nations. In 1967 devices to warn drivers of radar speed traps were being manufactured in the United Kingdom; they were deemed illegal under the Wireless Telegraphy Act 1949.

Radar Detector Detector

A radar detector detector (RDD) is a device used by police or law enforcement in areas where radar detectors are declared illegal.

Radar detectors are built around a superheterodyne receiver, which has a local oscillator that radiates slightly. It is therefore possible to build a radar-detector detector, which detects such emissions (usually the frequency of the radar type being detected, plus about 10 MHz for the intermediate frequency). Some radar guns are equipped with such a device.

However, like any device that detects stray emissions from electronic equipment, it is easily defeated by using adequate shielding.

History

The VG-2 Interceptor was the first device developed for this purpose, although more current technology such as the Spectre III (Stalcar in Australia) is now available. This form of "electronic warfare" cuts both ways and since detector-detectors use a similar superheterodyne receiver, many early "stealth" radar detectors were equipped with a radar-detector-detector-detector circuit, which shuts down the main radar receiver when the detector-detector's signal is detected, thus preventing detection by such equipment. In the early 1990s, BEL-Tronics, Inc. of Ontario, Canada (where radar detector use is prohibited) found that the local oscillator frequency of the detector could be altered to be out of the range of the VG-2 Interceptor. This resulted in a wave of detector manufacturers changing their local-oscillator frequency. Today, practically every radar detector on the market is immune to the VG-2 Interceptor. The VG-2 is no longer in production.

The Spectre III detected almost every radar detector certified for operation in the United States by the Federal Communications Commission as of December 2004. However counter technology has evolved rapidly, so that by July 2008, even budget radar detectors were able to avoid detection by the device. Then, in late 2008, the Spectre IV (Elite) was released, citing improved range and reliability over the Spectre III. Radar detector manufacturers produce some models undetectable by the Spectre Elite beyond a distance of a few inches, making them immune in real-world situations.

Radar Gun

Microdigicam Laser in use in Brazil

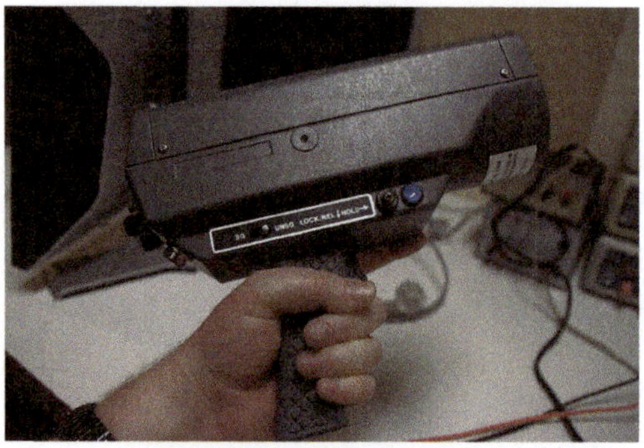

Handheld radar speed gun

A radar speed gun (also radar gun and speed gun) is a device used to measure the speed of moving objects. It is used in law-enforcement to measure the speed of moving vehicles and is often used in professional spectator sport, for things such as the measurement of bowling speeds in cricket, speed of pitched baseballs, athletes and tennis serves.

A radar speed gun is a Doppler radar unit that may be hand-held, vehicle-mounted or static. It measures the speed of the objects at which it is pointed by detecting a change in frequency of the returned radar signal caused by the Doppler effect, whereby the frequency of the returned signal is increased in proportion to the object's speed of approach if the object is approaching, and lowered if the object is receding. Such devices are frequently used for speed limit enforcement, although more modern LIDAR speed gun instruments, which use pulsed laser light instead of radar, began to replace radar guns during the first decade of the twenty-first century, because of limitations associated with small radar systems.

History

The radar speed gun was invented by John L. Barker Sr., and Ben Midlock, who developed radar for the military while working for the Automatic Signal Company (later Automatic Signal Division of LFE Corporation) in Norwalk, CT during World War II. Originally, Automatic Signal was approached by Grumman Aircraft Corporation to solve the specific problem of terrestrial landing gear damage on the now-legendary PBY Catalina amphibious aircraft. Barker and Midlock cobbled a Doppler radar unit from coffee cans soldered shut to make microwave resonators. The unit was installed at the end of the runway (at Grumman's Bethpage, NY facility), and aimed directly upward to measure the sink rate of landing PBYs. After the war, Barker and Midlock tested radar on the Merritt Parkway. In 1947, the system was tested by the Connecticut State Police in Glastonbury, Connecticut, initially for traffic surveys and issuing warnings to drivers for excessive speed. Starting in February 1949, the state police began to issue speeding tickets based on the speed recorded by the radar device. In 1948, radar was also used in Garden City, New York.

How it Works

U.S. Army soldier uses a radar speed gun to catch speeding violators at Tallil Air Base, Iraq.

Doppler Effect

Speed guns use Doppler radar to perform speed measurements.

Radar speed guns, like other types of radar, consist of a radio transmitter and receiver. They send out a radio signal in a narrow beam, then receive the same signal back after it bounces off the target object. Due to a phenomenon called the Doppler effect, if the object is moving toward or away from the gun, the frequency of the reflected radio waves when they come back is different from the transmitted waves. From that difference, the radar speed gun can calculate the speed of the object from which the waves have been bounced. This speed is given by the following equation:

$$v = \frac{\Delta f}{f} \frac{c}{2}$$

where c is the speed of light, f is the emitted frequency of the radio waves and Δf is the difference in frequency between the radio waves that are emitted and those received back by the gun. This equation holds precisely only when object speeds are low compared to that of light, but in everyday situations, this is the case and the velocity of an object is directly proportional to this difference in frequency.

Stationary Radar

After the returning waves are received, a signal with a frequency equal to this difference is created by mixing the received radio signal with a little of the transmitted signal. Just as two different musical notes played together create a beat note at the difference in frequency between them, so these two radio signals are mixed to create a "beat" signal (called a heterodyne) and an electrical circuit then measures this frequency using a digital counter and displays the number on a digital display as the object's speed.

Since this type of speed gun measures the difference in speed between a target and the gun itself, the gun must be stationary in order to give a correct reading. If a measurement is made from a moving car, it will give the difference in speed between the two vehicles, not the speed of the target relative to the road, so a different system has been designed to work from moving vehicles.

Moving Radar

In so-called "moving radar", a gun receives reflected signals from both the target vehicle and stationary background objects such as the road surface, nearby road signs, guard rails and streetlight poles. Instead of comparing the frequency of the signal reflected from the target with the transmitted signal, it compares the target signal with this background signal. The frequency difference between these two signals gives the true speed of the target vehicle.

Design Considerations

Modern radar speed guns normally operate at X, K, K_a, and (in Europe) K_u bands.

Radar guns that operate using the X band (8 to 12 GHz) frequency range are becoming less common because they produce a strong and easily detectable beam. Also, most automatic doors utilize radio waves in the X band range and can possibly affect the readings of police radar. As a result, K band (18 to 27 GHz) and K_a band (27 to 40 GHz) are most commonly used by police agencies.

Some motorists install radar detectors which can alert them to the presence of a speed trap ahead, and the microwave signals from radar may also change the quality of reception of AM and FM radio signals when tuned to a weak station. For these reasons, hand-held radar typically includes an on-off trigger and the radar is only turned on when the operator is about to make a measurement. Radar detectors are illegal in some areas.

Limitations

Traffic radar comes in many models. Hand-held units are mostly battery powered, and for the most part are used as stationary speed enforcement tools. Stationary radar can be mounted in police vehicles and may have one or two antennae. Moving radar is employed, as the name implies, when a police vehicle is in motion and can be very sophisticated, able to track vehicles approaching and receding, both in front of and behind the patrol vehicle and also able to track multiple targets at once. It can also track the fastest vehicle in the selected radar beam, front or rear.

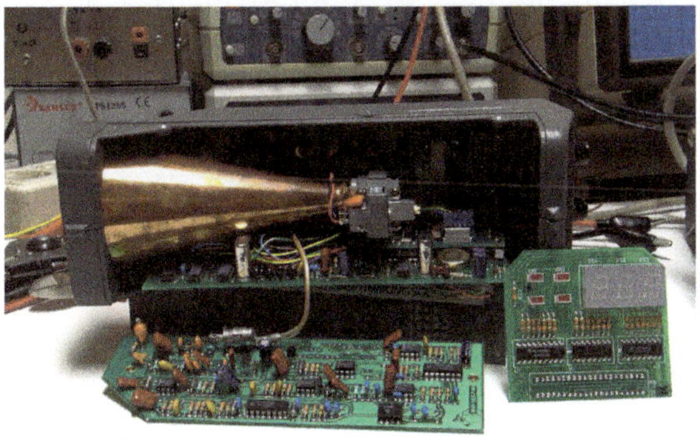

Disassembled radar speed gun. The copper cone is the microwave horn antenna.
At the right end is the Gunn diode oscillator which generates the microwaves.

However, there are a number of limitations to the use of radar speed guns. For example, user training and certification are required so that a radar operator can use the equipment effectively, with trainees being required to consistently visually estimate vehicle speed within +/-2 mph of actual target speed, for example if the targets actual speed is 30 mph than the operator must be able to consistently visually estimate the target speed as falling between 28 and 32 mph. Stationary traffic enforcement radar must occupy a location above or to the side of the road, so the user must understand trigonometry to accurately estimate vehicle speed as the direction changes while a single vehicle moves within the field of view. Actual vehicle speed and radar measurement thus are rarely the same, however, for all practical purposes this difference in actual speed and measured speed is inconsequential, generally being less than 1 mph difference, as police are trained to position the radar to minimize this inaccuracy and when present the error is always in the favor of the driver reporting a lower than actual speed. Radar speed guns do not differentiate between targets in traffic, and proper operator training is essential for accurate speed enforcement. This inability to differentiate among targets in the radars field of view is the primary reason for the operator being required to consistently and accurately visually estimate target speeds to within +/-2 mph, so that, for example if there are seven targets in the radars field of view and the operator is able to visually estimate the speed of six of those targets as approximately 40 mph and visually estimate the speed of one of those targets as approximately 55 mph and the radar unit is displaying a reading of 56 mph it becomes clear which targets speed the unit is measuring.

Size

The primary limitation of hand held and mobile radar devices is size. An antenna diameter of less than several feet limits directionality, which can only partly be compensated for by increasing the frequency of the wave. Size limitations can cause hand-held and mobile radar devices to produce measurements from multiple objects within the field of view of the user.

The antenna on some of the most common hand-held devices is only 2 inches (5.1 cm) in diameter. The beam of energy produced by an antenna of this size using X-band frequencies occupies a cone that extends about 22 degrees surrounding the line of sight, 44 degrees in total width. This beam is called the main lobe. There is also a side lobe extending from 22 to 66 degrees away from the line of sight, and other lobes as well, but side lobes are about 20 times less sensitive than the main lobe (13 dB), although they will detect moving objects close by. The primary field of view is about 130 degrees wide. K-band reduces this field of view to about 65 degrees by increasing the frequency of the wave. Ka-band reduces this further to about 40 degrees. Side lobe detections can be eliminated using side lobe blanking which narrows the field of view, but the additional antennas and complex circuitry impose price constraints that limit this to applications for the military, air traffic control, and weather agencies. Mobile weather radar is mounted on semi-trailer trucks in order to narrow the beam.

Distance

A second limitation for hand-held devices is that they have to use continuous-wave radar to make them light enough to be mobile. Speed measurements are only reliable when the distance at which a specific measurement has been recorded is known. Distance measurements require pulsed operation or cameras when more than one moving object is within the field of view. Continuous-wave radar may be aimed directly at a vehicle 100 yards away but produce a speed measurement from a second vehicle 1 mile away when pointed down a straight roadway. Once again falling back on the training and certification requirement for consistent and accurate visual estimation so that operators can be certain which object's speed the device has measured without distance information, which is unavailable with continuous wave radar.

Some sophisticated devices may produce different speed measurements from multiple objects within the field of view. This is used to allow the speed-gun to be used from a moving vehicle, where a moving and a stationary object must be targeted simultaneously, and some of the most sophisticated units are capable of displaying up to four separate target speeds while operating in moving mode once again emphasizing the importance of the operators' ability to consistently and accurately visually estimate speed.

Environment

The environment and locality in which a measurement is taken can also play a role. Using a hand-held radar to scan traffic on an empty road while standing in the shade of a large tree, for example, might risk detecting the motion of the leaves and branches if the wind is blowing hard (side lobe detection). There may be an unnoticed airplane overhead, particularly if there is an airport nearby. Again emphasize the importance of proper operator training.

Associated Cameras

Conventional radar gun limitations can be corrected with a camera aimed along the line of sight.

Cameras are associated with automated ticketing machines (known in the UK as speed cameras) where the radar is used to trigger a camera. The radar speed threshold is set at or above the maximum legal vehicle speed. The radar triggers the camera to take several pictures when a nearby object exceeds this speed. Two pictures are required to determine vehicle speed using roadway survey markings. This can be reliable for traffic in city environments when multiple moving objects are within the field of view. It is the camera, however, and its timing information, in this case, that determines the speed of an individual vehicle, the radar gun simply alerting the camera to start recording.

Newer Instruments

Laser devices, such as a LIDAR speed gun, are capable of producing reliable range and speed measurements in typical urban and suburban traffic environments without the site survey limitation and cameras. This is reliable in city traffic because LIDAR has directionality similar to a typical firearm because the beam is shaped more like a pencil that produces measurement only from the object it has been aimed at.

In Media

MythBusters did an episode on trying to get the gun to have incorrect readings by changing the surface of the passing object.

Radar Warning Receiver

Radar warning receiver (RWR) systems detect the radio emissions of radar systems. Their primary purpose is to issue a warning when a radar signal that might be a threat (such as a police speed detection radar) is detected. The warning can then be used, manually or automatically, to evade the detected threat. RWR systems can be installed in all kind of airborne, sea-based, and ground-based assets (such as aircraft, ships, automobiles, military bases). This section is focused mainly on airborne military RWR systems; for commercial police RWR systems.

Depending on the market the RWR system is designed for, it can be as simple as detecting the presence of energy in a specific radar band (such as police radar detectors). For more critical situations, such as military combat, RWR systems are often capable of classifying the source of the radar by the signal's strength, phase and waveform type, such as pulsed power wave or continuous wave with amplitude modulation or frequency modulation (chirped). The information about the signal's strength and waveform can then be used to estimate the most probable type of threat the detected radar poses. Simpler systems are typically installed in less expensive assets like automobiles, while more sophisticated systems are installed in mission critical assets such as military aircraft.

Description

The RWR usually has a visual display somewhere prominent in the cockpit (in some modern aircraft, in multiple locations in the cockpit) and also generates audible tones which feed into the pilot's (and perhaps RIO/co-pilot/GIB's in a multi-seat aircraft) headset. The visual display often takes the form of a circle, with symbols displaying the detected radars according to their direction relative to the current aircraft heading (i.e. a radar straight ahead displayed at the top of the circle, directly behind at the bottom, etc.). The distance from the center of the circle, depending on the type of unit, can represent the estimated distance from the generating radar, or to categorize the severity of threats to the aircraft, with tracking radars placed closer to the center than search radars. The symbol itself is related to the type of radar or the type of vehicle that carries it, often with a distinction made between ground-based radars and airborne radars.

The typical airborne RWR system consists of multiple wideband antennas placed around the aircraft which receive the radar signals. The receiver periodically scans across the frequency band and determines various parameters of the received signals, like frequency, signal shape, direction of arrival, pulse repetition frequency, etc. By using these measurements, the signals are first deinterleaved to sort the mixture of incoming signals by emitter type. These data are then further sorted by threat priority and displayed.

The RWR is used for identifying, avoiding, evading or engaging threats. For example, a fighter aircraft on a combat air patrol (CAP) might notice enemy fighters on the RWR and subsequently use its own radar set to find and eventually engage the threat. In addition, the RWR helps identify and classify threats—it's hard to tell which blips on a radar console-screen are dangerous, but since different fighter aircraft typically have different types of radar sets, once they turn them on and point them near the aircraft in question it may be able to tell, by the direction and strength of the signal, which of the blips is which type of fighter.

A non-combat aircraft, or one attempting to avoid engagements, might turn its own radar off and attempt to steer around threats detected on the RWR. Especially at high altitude (more than 30,000 feet AGL), very few threats exist that don't emit radiation. As long as the pilot is careful to check for aircraft that might try to sneak up without radar, say with the assistance of AWACS or GCI, it should be able to steer clear of SAMs, fighter aircraft and high altitude, radar-directed AAA.

SEAD and ELINT aircraft often have sensitive and sophisticated RWR equipment like the U.S. HTS (HARM targeting system) pod which is able to find and classify threats which are much further away than those detected by a typical RWR, and may be able to overlay threat circles on a map in the aircraft's multi-function display (MFD), providing much better information for avoiding or engaging threats, and may even store information to be analyzed later or transmitted to the ground to help the commanders plan future missions.

The RWR can be an important tool for evading threats if avoidance has failed. For example, if a SAM system or enemy fighter aircraft has fired a missile (for example, a SARH-guided missile) at the aircraft, the RWR may be able to detect the change in mode that the radar must use to guide the missile and notify the pilot with much more insistent warning tones and flashing, bracketed symbols on the RWR display. The pilot then can take evasive action to break the missile lock-on or dodge the missile. The pilot may even be able to visually acquire the missile after being alerted to the possible launch. What's more, if an actively guided missile is tracking the aircraft, the pilot can use the direction and distance display of the RWR to work out which evasive maneuvers to perform to outrun or dodge the missile. For example, the rate of closure and aspect of the incoming missile may allow the pilot to determine that if they dive away from the missile, it is unlikely to catch up, or if it is closing fast, that it is time to jettison external supplies and turn toward the missile in an attempt to out-turn it. The RWR may be able to send a signal to another defensive system on board the aircraft, such as a Countermeasure Dispensing System (CMDS), which can eject countermeasures such as chaff, to aid in avoidance.

RWR Types in Service

- ALR-2002 (Australia; developmental, project being cancelled)

- ALR-400 (Spain; EF-18A/B Hornet, Airbus A400M, C-295, CH-47 Chinook, Cougar, TIGER, NH90, CH-53)
- AN/APR-39 (USA, AH-1, AH-64 Apache, CH-46 Sea Knight, CH-47 Chinook, CH-53, EH-60 Black Hawk, KC-130 Hercules, MH-47 Chinook, MH-60 Black Hawk, OH-58, OV-1 Mohawk, RC-12, RV-1, UH-1 Iroquois, UH-60 Black Hawk, V-22 Osprey)
- AN/ALR-46 (F-4 Phantom II, RF-4 Phantom II, F-5, B-52G)
- AN/ALR-56 (USA; F-15 Eagle, F-16 block 50 and 52, Canada; CC-130 Hercules)
- AN/ALR-66 (USA; P-3C Orion)
- AN/ALR-67 (USA; AV-8B Harrier II, F-14 Tomcat, F/A-18 Hornet, F/A-18E/F Super Hornet, EA-6B Prowler, Canada; CF-18 Hornet)
- AN/ALR-69 (USA; B-52H Stratofortress, lacks Frequency Selective Receiver Capabilities (FSRS) capabilities however, A-10 Thunderbolt II, AC-130 Spectre, F-16 Fighting Falcon, HH-53, MC-130, F-4E Phantom II).
- AN/ALR-76 (USA; S-3, EP-3)
- AN/ALR-94 (USA; F-22)
- ARI 18223 (British aircraft such the Jaguar Mk.1)
- BOW-21 (Sweden; JAS 39 Gripen Germany; Panavia Tornado)
- L-150 Pastel (Soviet-Russian aircraft)
- Sirena series (Soviet aircraft)
- SkyGuardian 2000 (British aircraft as EH-101 and Tornado; Portuguese EH-101)
- SPO-15 Beryoza (India; MiG-29 Fulcrum, Russia; Su-27SK Flanker-B, China; J-11)
- SPS-1000V5 (Portugal; F-16 Fighting Falcon, C-295M)
- Tarang (India; MiG-27 Flogger, LCA Tejas, Jaguar, Su-30MKI Flanker)

Precision Approach Radar

Precision approach radar (PAR) is a type of radar guidance system designed to provide lateral and vertical guidance to an aircraft pilot for landing, until the landing threshold is reached. After the aircraft reaches the decision height (DH) or decision altitude (DA), guidance is advisory only. Controllers monitoring the PAR displays observe each aircraft's position and issue instructions to the pilot that keep the aircraft on course and glidepath during final approach. It is similar to an instrument landing system (ILS) but requires control instructions. One type of instrument approach that can make use of PAR is the ground-controlled approach (GCA). Air traffic controllers must transmit a minimum of every 5 seconds to the pilot their relation to the azimuth portion and, once intercepting the glidepath, their elevation. The approach is terminated when the aircraft reaches the OCA/H (Obstacle Clearance Altitude/Height). Nevertheless, information is provided till threshold and aircraft may be monitored by controller till touchdown. Controller in charge of PAR should not be responsible for any duty other than the PAR approach concerned.

Precision approach radar PAR-80 on a military airfield in Germany

The upper portion of the display indicates elevation, the lower portion azimuth. Controllers must be able to interpret radar returns for the azimuth as a "top view" to inform them if the aircraft is left or right of course.

An Air Force air traffic controller is reflected in the precision approach radar scope (1980)

Precision approach radars are most frequently used at military air traffic control facilities. Many of these facilities use the AN/FPN-63, AN/MPN, or AN/TPN-22. These radars can provide precision guidance to a distance of 10 to 20 miles. The OJ-333 Radar scope is the indicator which the air traffic controller uses to provide instructions to the pilots.

In the United States PAR are used mostly by The Navy. This is because they present a more covert type of precision approach for use on Aircraft carriers. An ILS installed on a ship could provide guidance to enemy missiles but a PAR does not provide accurate guidance without controller instruction.

Non-traditional PAR using SSR Transponder Reply

There are systems that provide PAR functionality without using primary radar. These non-traditional PAR systems use transponder multilateration, triangulation and/or trilateration.

One such system, Transponder Landing System (TLS) precisely tracks aircraft using the mode 3/A transponder response received by antenna arrays located near the runway. These antennas are part of a measurement subsystem that is used to precisely determine the aircraft 3-dimensional position using TOA, DTOA and AOA measurement techniques. The aircraft position is then displayed on a high-resolution color graphics terminal that also shows the approach centerline and the glide path. A GCA controller is then able to use this screen for reference to issue GCA instructions to the pilot.

The signal strength for the secondary surveillance radar subsystem of a non-traditional PAR is not attenuated by rain since the frequency is within the long range band, L-band. Therefore, a non-traditional PAR does not experience noticeable rain fade and in the case of the TLS has an operational range of 60 nm.

This system is co-operative depending, it means that in the case of transponder failure no aircraft detection will be provided.

Flight Inspection of the PAR

A traditional PAR flight inspection procedure is performed without a navigation signal available to compare directly to a truth reference. A traditional PAR is flight inspected by comparing written notes between two observers, one taking notes at a truth reference system such as a theodolite and the other observer taking notes while observing the radar console. The Transponder Landing System (TLS) non-traditional PAR can transmit an ILS signal that corresponds to the aircraft position relative to the precision approach. Therefore, the graphical depiction can be directly verified using Instrument Landing System (ILS) flight inspection techniques. This direct measurement removes some ambiguity from the PAR flight inspection process.

Terrain-following Radar

Terrain-following radar (TFR) is an aerospace technology that allows a very-low-flying aircraft to automatically maintain a relatively constant altitude above ground level. It is sometimes referred-to as *ground hugging* or *terrain hugging* flight. The term *nap-of-the-earth* flight may also apply but is more commonly used in relation to low-flying military helicopters, which typically do not use terrain-following radar. The technology was originally developed by Ferranti for use with the TSR-2 aircraft.

Technology and usage

The system works by transmitting a radar signal towards the ground area in front of the aircraft. The radar returns can then be analysed to see how the terrain ahead varies, which can then be used by the aircraft's autopilot to maintain a reasonably constant height above the earth.

TSR-2 *XR220* at RAF Museum Cosford, 2002. Ferranti developed terrain following radar specifically for the TSR-2.

This technology is primarily used by military strike aircraft, to enable flight at very low altitudes (sometimes below 100 feet (30 metres)) and high speeds, avoiding detection by enemy radars and interception by anti-aircraft systems. Normally, this radar is used by the navigator and allows the pilot to focus on other aspects of the flight besides the extremely intensive task of low flying itself. It can also enable low-altitude flight at night and in other low-visibility conditions.

Some aircraft such as the Tornado IDS have two separate radars, with the smaller one used for terrain-following. However more modern aircraft such as the Rafale with phased array radars can look forward and at the ground simultaneously, by electronically steering one or more beams in milliseconds.

Most aircraft allow the pilot to select the ride "hardness", to choose between how closely the aircraft tries to keep itself close to the ground and the forces exerted on the pilot. The F-111 used a switch to select for a hard, medium or a soft ride.

The TFR computer will consider many factors in determining the flight path for the aircraft. These factors include, distance to the forward terrain, aircraft speed and velocity, angle of attack and quality of signal being returned.

The F-111C employs TFR

Terrain-following radar is also sometimes used by civilian aircraft that map the ground and wish to maintain a constant height over it.

Military helicopters may also have terrain-following radar. Due to their lower speed and high maneuverability, helicopters are normally able to fly lower than fixed wing aircraft.

Advantages and Disadvantages

By flying at very low altitude, the aircraft can take advantage of terrain masking and avoid detection by enemy radar systems. The system is also largely automated and can take some of the workload off the pilot.

The radar emissions can be detected by enemy anti-aircraft systems with relative ease once there is no covering terrain, allowing the aircraft to be targeted. The use of terrain-following radar is therefore a compromise between the increased survivability due to terrain-masking and the ease with which the aircraft can be targeted if it is seen.

Since the radar cannot tell what is beyond any immediate terrain, the flight path may suffer from "ballooning" over the top of high ground. Furthermore, small obstacles such as radio antennas and electricity pylons may not show up on the radar at all. The objects present collision hazards when flying at very low level.

Even an automated system has limitations, and all aircraft with terrain-following radars installed have limits on how low and fast they can fly. Factors such as system response-time, aircraft g-limits and the weather can all limit an aircraft.

Alternatives

There are very few alternatives to using terrain-following radar for high-speed, low altitude flight. TERPROM, a terrain-referenced navigation system provides a limited but passive terrain-following functionality.

References

- Martial, Hebert (1992). "3-D Measurements From Imaging Laser Radars: How Good Are They?". International Journal of Image and Vision Computing. 10 (3): 170–178
- V. Tuchin (2000). Tissue Optics Light Scattering Methods and Instruments for Medical Diagnosis. Society of Photo Optical. ISBN 0-8194-3459-0
- Hofinghoff, Jan-Florian (2013). "Resistive Loaded Antenna for Ground Penetrating Radar Inside a Bottom Hole Assembly". IEEE Transactions on Antennas and Propagation. IEEE. 61 (12). doi:10.1109/TAP.2013.2283604
- Mallik, Udayan (2010). "A 3D IMAGING RADAR A Monograph on a 256 Receiver Antenna, 3-D, Multi-wave and Multi-band, Imaging RADAR". Technical Journal of Udayaninc: 3
- MIT Lincoln Laboratory (2016-06-24), Enabling Autonomous Vehicles to Drive in the Snow with Localizing Ground Penetrating Radar, retrieved 2017-05-31
- Daniels DJ (ed.) (2004). Ground Penetrating Radar (2nd ed.). Knoval (Institution of Engineering and Technology). pp. 1–4. ISBN 978-0-86341-360-5
- N Masini, R Persico, E Rizzo. Some examples of GPR prospecting for monitoring of the monumental heritage. Journal of Geophysics and Engineering 7 (2), 190, doi:10.1088/1742-2132/7/2/S05
- Aftanas, Michal (2010). Through-Wall Imaging With UWB Radar System (PDF). Berlin: LAP LAMBERT Academic Publishing. p. 132. ISBN 3838391764
- "MIT Lincoln Laboratory: News: Lincoln Laboratory demonstrates highly accurate vehicle localization under adverse weather conditions". www.ll.mit.edu. Retrieved 2017-05-31

Permissions

All chapters in this book are published with permission under the Creative Commons Attribution Share Alike License or equivalent. Every chapter published in this book has been scrutinized by our experts. Their significance has been extensively debated. The topics covered herein carry significant information for a comprehensive understanding. They may even be implemented as practical applications or may be referred to as a beginning point for further studies.

We would like to thank the editorial team for lending their expertise to make the book truly unique. They have played a crucial role in the development of this book. Without their invaluable contributions this book wouldn't have been possible. They have made vital efforts to compile up to date information on the varied aspects of this subject to make this book a valuable addition to the collection of many professionals and students.

This book was conceptualized with the vision of imparting up-to-date and integrated information in this field. To ensure the same, a matchless editorial board was set up. Every individual on the board went through rigorous rounds of assessment to prove their worth. After which they invested a large part of their time researching and compiling the most relevant data for our readers.

The editorial board has been involved in producing this book since its inception. They have spent rigorous hours researching and exploring the diverse topics which have resulted in the successful publishing of this book. They have passed on their knowledge of decades through this book. To expedite this challenging task, the publisher supported the team at every step. A small team of assistant editors was also appointed to further simplify the editing procedure and attain best results for the readers.

Apart from the editorial board, the designing team has also invested a significant amount of their time in understanding the subject and creating the most relevant covers. They scrutinized every image to scout for the most suitable representation of the subject and create an appropriate cover for the book.

The publishing team has been an ardent support to the editorial, designing and production team. Their endless efforts to recruit the best for this project, has resulted in the accomplishment of this book. They are a veteran in the field of academics and their pool of knowledge is as vast as their experience in printing. Their expertise and guidance has proved useful at every step. Their uncompromising quality standards have made this book an exceptional effort. Their encouragement from time to time has been an inspiration for everyone.

The publisher and the editorial board hope that this book will prove to be a valuable piece of knowledge for students, practitioners and scholars across the globe.

Index

R

S

T

U

V

W